Air-Mech
3-Dimensional Phalanx

**FULL SPECTRUM MANEUVER
WARFARE TO DOMINATE
THE 21ST CENTURY**

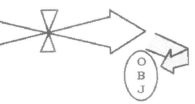

**Brigadier General David L. Grange (Retired)
Brigadier General Huba Wass de Czege (Retired)
Lieutenant Colonel Richard D. Liebert, USAR
Major Charles A. Jarnot, U.S. Army
Mike Sparks, USARNG**

**Foreword by LTG Harold G. "Hal" Moore (Retired)
Best-selling Author of "We Were Soldiers Once... and Young"**

Air-Mech-Strike
3-DIMENSIONAL *PHALANX*

Full-spectrum Maneuver Warfare for the 21st Century

BG DAVID L. GRANGE, U.S. Army (Retired)
BG HUBA WASS DE CZEGE, U.S. Army (Retired)
LTC RICHARD D. LIEBERT, USAR
MAJ CHARLES A. JARNOT, U.S. Army
MICHAEL L. SPARKS, USARNG
Foreword by LTG Harold G. Moore U.S. Army (Retired)

Copyright © 2000 by the *Air*-Mech-Strike Study Group (Airborne)
Publishing Rights: Turner Publishing Company

All rights reserved. No portion of this book may be reproduced, by a process or technique, without the express written consent of the *Air*-Mech-Strike Study Group (Airborne) or the publisher.

Library of Congress Catalog Card Number: 00-105946
ISBN: 1-56311-616-2

First published in 2000
Turner Publishing Company
P.O. Box 3101, Paducah, KY 42002-3101
(270) 443-0121

Printed in the United States of America

NOTICE: The opinions and scenarios expressed in this book are those of the authors only and do not express the policies or plans of the U.S. Department of Defense or the U.S. Army.

Air-Mech-Strike
3-DIMENSIONAL *PHALANX*

Rwanda? Kosovo? Bosnia? Iraq? Korea? What next?
A conflict has broken out in some little known corner of the world in the early part of the 21st Century. Deep behind forbidding mountainous terrain, the world watches a war unfold. The enemy is hiding behind civilians and restricted terrain making them unreachable by "precision-guided munitions" and air strikes. CNN is commenting each hour about the continuing atrocities, and NATO airstrike damage to the civilian infrastructure. NATO is deadlocked about what to do next. They conclude the only solution is a ground operation.

America's answer?
U.S. Army 3-Dimensional *Air*-Mech-Strike Forces.

This book outlines how to reorganize the U.S. Army into a fully 3-dimensional maneuver, ground forces with terrain-agile, armored fighting vehicles sized to rapidly deploy by fixed-wing and rotary-wing aircraft to the scene of world conflicts and strike at the heart of freedom's enemies. The plan to build the Army into *Air*-Mech-Strike Forces, exploiting emerging information-age technologies, as well as America's supremacy in aircraft and helicopter delivery systems, at the lowest cost to the taxpayers, is described in detail. These Army warfighting organizations, using existing and some newly purchased equipment, will shape the battlefield to America's advantage, preserving the peace before it is lost; if not, then winning fights that must be fought quickly. The dangerous world we live in moves by the speed of the AIR, and the 21st Century U.S. Army *Phalanx* will dominate this medium by *Air*-Mech-Strike!

Dedication

This book is dedicated to **the present and future U.S. Army Soldier** who will have to win the wars and maintain the peace in an increasingly dangerous world. To help him in this duty and to give us direction, we also dedicate this book to the memory of two of the U.S. Army's and America's—greatest heroes.

Lieutenant General James Maurice Gavin, whose concept of an Airborne Armored U.S. Army force that can fly into battle and have maximum terrain agility, will be fulfilled in the 21st Century. A man ahead of his time who "wrote the book" on how U.S. Airborne forces would drop into battle ready to fight not have to look for separate equipment containers. A man who was always the first to jump into battle in WWII. After the war, he wrote "the book" on Airborne warfare, got us our first air-deployed armored personnel carriers, light tanks, improved parachuting and promoted the Air Assault concept used successfully during Vietnam. We honor him by naming the Medium Armored Vehicle of the immediate future —the "Gavin Fighting Vehicle".

General Matthew Bunker Ridgway, who was a senior Airborne commander during WWII. A leader whose firepower tactics turned the tide of battle during the bleak hours of the Korean War. A man, who had already served decades in peace and war, but walked the line of our frostbitten troops seeing to it that they were taken care of with boots, socks and warm food, yet had grenades taped to his web gear and rifle in hand ready to fight. We name the future helo-deployable Reconnaissance/*Air* Assault armored vehicle of the U.S. Army the "*Ridgway*" after him for his love of Soldiers and his contributions towards 3-dimensional warfare.

THE STAFF, *AIR*-MECH-STRIKE! 3-DIMENSIONAL *PHALANX*
Brigadier General David L. Grange, Retired, U.S. Army
Brigadier General Huba Wass de Czege, Retired, U.S. Army
Lieutenant Colonel Richard D. Liebert , U.S. Army Reserve
Lieutenant Colonel Lester Grau, Retired, U.S. Army
Major Charles A. Jarnot, U.S. Army
Emery E. Nelson   Michael L. Sparks
Jacob W. Kipp   John Richards
Carol Murphy

Contents

Illustrations 6
Acknowledgments 7
Foreword by Lieutenant General Harold G. Moore (Ret.) 12

1. Introduction 15
2. Purpose 25
3. Understanding U.S. Army 3-D maneuver history 34
   Russian Airborne 3-D *Air*-Mech maneuver in combat 54
   Jacob W. Kipp & Lester W. Grau
4. *Air*-Mech-Strike: 3-D Army maneuver warfare theory 76
5. The 12 *Air*-Mech-Strike Axioms 106
6. *Air*-Mech-Strike is possible today 117
7. 3-D *Air*-Mech-Strike weapons, equipment, ground vehicles
   and C4 ISR 135
8. 3-D *Air*-Mech-Strike organizations 186
9. *Air*-Mech-Strike *Phalanx* in action 215
   Interim Air-Mech Force in Kosovo 219
   Interim Air-Mech Force in Korea 220
   Interim Air-Mech Force in Southwest Asia 225
10. 3-D *Air*-Mech-Strike training and readiness 232
11. The U.S. Army's *Air*-Mechanised future 238
    Wing-In-Ground effect aircraft 252
    Jacob W. Kipp & Lester W. Grau
12. Summary/Conclusions 276

Bibliography 284
Appendix A: Glossary of Military Terms 292
Appendix B: Air-Mech-Strike Illustrated 295
Index 305
Staff Biographies 308

Illustrations

Figure 1	Soldier Poem	11
Figure 2	British CH-47s "*Air*-Meching" vehicles into Kosovo	18
Figure 3	4-Ton RFV Strategic 747 Air Mobility frees C-17 Airlift for Heavy Combat Modules	33
Figure 4	How *Air*-Mech-Strike builds bridge to the future	57
Figure 5	Russian ASU-57 Assault guns	57
Figure 6	Proposed *Ridgway* Fighting Vehicle (RFV)	75
Figure 7	Proposed AMS *Gavin* Fighting Vehicle (GFV)	102
Figure 8	UH-60 Streamlined External Load (SEL) of RFV	103
Figure 9	Rotary-wing Air Delivery methods	104
Figure 10	Airdrop: Low-Velocity parachutes (Para-Mech)	105
Figure 11	Airland: Mobile Gun System rolling off C-130 *Hercules*	115
Figure 12	Air-Mech Vehicle superior 2-D agility	139
Figure 13	Flyer 21 ATV	177
Figure 14	*Air*-Mech-Strike AFV	185
Figure 15	M* Armored Gun Systems (AGS)	195
Figure 16	*Air*-Mech-Strike symbology	196
Figure 17	Army Combat Division: HEAVY chart	207
Figure 18	Army Combat Division: LIGHT chart	208
Figure 19	Army Combat Division: AIR ASSAULT chart	209
Figure 20	Army Combat Division: AIRBORNE chart	210
Figure 21	*Gavin* Fighting Vehicle Infantry Battalion	211
Figure 22	*Ridgway* Fighting Vehicle Infantry Battalion	212
Figure 23	*Ridgway* 120mm mortar; roof firing and turreted	213
Figure 24	Kosovo situation map: What if we had attacked in 1999?	214
Figure 25	Korean situation map: DEFENSIVE operations 2003	228
Figure 26	Korean situation map: OFFENSIVE operations 2003	229
Figure 27	Kuwait situation map: DETERRENCE operations 2003	230
Figure 28	Russian Wing-In-Ground (WIG) effect aircraft	231
Figure 29	Conflict Resolution SO/Combat with *Air*-Mech-Strike	267

Acknowledgments

We acknowledge the following people for their hard work in making this book possible: to **Colonel Douglas MacGregor** by publishing the first *Phalanx* book and sending us his briefing to the Army Chief of Staff; he explained how Air-Mechanized maneuvers are being extensively studied/wargamed by both U.S. and its allies for future military operations. One of those leading advocates of American *Air*-Mech theory, **Brigadier General Huba Wass de Czege (Retired)** has been very inspiring to us with his many writings on the subject. **General Wayne Downing (Retired)** provided extremely valuable reviews of the early rough drafts of the book. **Mr. Jacob W. Kipp** and **Lieutenant Colonel Lester W. Grau (Retired)** provided an amazing look into the Russian Airborne's *Air*-Mech past and into the future of *Air*-Mech aircraft transporting M1/M2 heavy AFVs, superbly illustrated by Long-Range Reconnaissance Surveillance Detachment Paratrooper **Mr. John Richards**.

We owe a tremendous debt to **Ms. Carol Murphy** who typed large parts of the manuscript and reformatted it for submission, researching hard-to-get facts and documents and generally spending countless hours getting it right.

We thank USAF Intelligence **Captain Eric Graves** for his help collecting open source-information from the word wide web. Vietnam Intelligence analyst **Roy S. Ardillo** for his insights into 3-D maneuver during the battle for An Loc, Air Defense Artillery Veteran **Red Sigle** for the outstanding pictures and information of truck-mounted Quad .50 caliber machine guns and CH-47 sling-loads used for early air-portable firepower; **Mr. Christopher Van Buiten,** Chief of Preliminary Design and his able fellow engineer, **Mr. Keith McVicar** of Sikorsky Aircraft for their creative options to maximize *Air-Mech-Strike* potential. **Mr. Dennis McCarthy** of TACOM for the accurate info on the state of the M113A3 AFV fleet in the U.S. Army. **Mrs. Holly (Getz) Grange** for her family's hospitality to the book staff and her insights into the Army from her own experiences as an Army engineer officer. **Lieutenant Colonel Michael D. Ryan** for his strong support editing Army aviation concepts. **Colonel George S. Webb** for his pioneering monograph on *Air*-Mechanization 1986 for the U.S. Army School of Advanced Military Studies, Fort Leavenworth, Kansas. Engineers **Bill Grusonik** and **Val Horvatich** of United Defense gave us valuable information on the M113A3's precise weights for helicopter transportability and state-of-the-art in lightweight titanium appliqué armor capable of being applied to armored vehicles to defeat 30mm autocannon and RPG weaponry. **Retired LTC John B. Whitehead III**, 11th ACR and decorated Vietnam war helicopter pilot provided invaluable help with operational concepts and detailed information on the M113A3 and M8 Armored Gun System armored vehicles. **Mrs. Paula J. Barczewski Jarnot,** for her unfailing support and editorial advice. **CW4 John Leak** and **CW4 David McMahon** for their advice on CH-47 performance capabilities. Major Chuck Wright, **CW3 Edwin Watson** and **CW3 Craig Richardson** for their advice on UH-60 performance capabilities. **MSG John Morris**, **MSG Patrick Troxel** and **SFC Marcus Brown** on their flight operations and sling loading advice.

The Republic of Singapore Air Force **Lieutenant Colonel Dennis Phua** and **Major Jerome Wong**, for their outstanding *Chinook* sling-loading M113s photographs. **Major Dave Daigle** and **Mr. Jon Clemons** of U.S. Army Armor magazine for their information on *Sheridan* light tanks being airdropped into combat in Panama and the fine picture of a CH-47C *Chinook* sling-loading a M113A1 from Vietnam. We thank Vietnam tanker combat veteran-turned-historian **Ralph Zumbro** for his excellent writings of 3-Dimensional armored warfare. We cite former **Marine Captain Don Loughlin** for his inspired writings defending the Armored Fighting Vehicle industrial base of America. The extraordinary writings of defense analyst, **Scott Gourley** have helped us immensely with defining the issues involved and are the tools we used to fully realize the potential of existing Army vehicles. **Retired combat veteran Colonel Bruce B.G. Clarke** is commended for sharing his vision with us of a terrain-agile, situationally aware force that can disperse to evade enemy targeting/fires yet has a large caliber gun system to dominate the close fight in addition to skillfully employing joint fires. Vietnam M113 ACAV combat veteran **Bruce O'Keefe** for his insights into Airborne mechanized units of the

1/2d Battalions of the 509th with the 8th Infantry Division in Germany. **Bob Novogratz** and retired **General Volney Warner** got us accurate data on the German *Air*-Mech model and provided senior mentorship on the need for 3-D maneuver. The German Embassy military science advisor, **Mr. Hans-Georg Bergmann** provided detailed data on the German Army's *Wiesel* family of *Air*-Mech vehicles to include exact fuel consumption figures. **Miss Kim Burger** of *Inside the Army* is commended for her courageous writings on the truth about the Army BCT vehicle selection process. **Mr. John Pike's** Federation of American Scientist's military equipment www pages were invaluable aids for analysis and are world-class. **Mr. Thomas White** for his legal advice and his excellent military history perspectives. **Dr. Sherwood S. Cordier** for his mentorship and professional assistance in the early development of the original "*Rotor-Blitz*" idea to the "*Air-Mech-Strike*" concept of today. **Retired Major Tracy Pittman** for his strategic deployment advice. **Retired Lieutenant Colonel James Moon** for his contingency force projection advice. **Lieutenants Colonels Chester Kojro** provided fantastic and historical insights into the nature of mobile and urban warfare, which we all used to define the debate, which **Emery Nelson** used to spearhead the chapter on history. **Colonel John F. Antal III** provided excellent ideas on how the battlefield is actually as Sun Tzu depicted it—with "ordinary" 2-D and "extraordinary" 3-D forces moving in concert; that it can be seen as either a close or far fight and to accelerate decision making to execute *Air*-Mech warfare. **Colonel Dan Bolger** helped with advice and encouragement and his own valuable insights from his many books. **Major Don Vandegriff** inspired us with his work on digitization and the personnel management system and how both must be improved to fully exploit *Air*-Mech-Strike. **Sergeant First Class Ernest Hoppe**, Special Forces Medic and Engineer did much of the early detective work figuring out the actual weights of Army equipment and parachuted with **1LT Jeff Schram**, **Michael Sparks**, retired SF **Master Sergeant Lee Cashwell**, **Skip Somers** and Military Free-Fall Instructor **Dave Lilico** during the Operation *Dark Claw* human-powered vehicle mobility demonstration for Fort Bragg SF Combat developers. **SGT**

Andy Curtis, 18C Special Forces Combat Engineer related his views on the type of *Air*-Mech armored vehicle special operations could employ.

U.S. Army National Guardsmen including **Staff Sergeant Richard Kelly** gave us a guided tour of the capabilities of the M1 *Abrams* series tank, 197th Infantry Brigade combat veterans **Sergeants Elliot Ingram, John Miller and Vincent Garrett** described how they used M113 series vehicles to lead the way into Iraq with their cross-country mobility for Desert Storm. **Sergeant Mike Currier** related his Operation Just Cause experiences in Panama as a M1064 mortar M113 Soldier. **Staff Sergeant Larry Garrett** explained how Electro-Optical CounterMeasures work and could be applied to other platforms. **Sergeant First Class Travis Rigsby** related his observations on how a light and a mechanized infantry unit differ and the impacts the M113 and BFV-type vehicle has on unit outlook.

We thank **retired Colonel Douglas Coffey** for righting the public debate for the Army's new armored vehicle. We applaud the work of **retired Lieutenant General "Hal" Moore** fighting for a 3-Dimensional maneuver-capable U.S. Army. Lastly, we salute the serving U.S. Army Soldier today, in the active Army, the Army National Guard and Army Reserves. You are NOT forgotten and in the interests of your country and the cause of freedom we offer this work to give you the tools to be successful on the modern battlefield. We especially thank Vietnam combat veteran, military historian and acclaimed artist, **Captain George L. Skypeck** for allowing us to use his famous poem, beautifully illustrated—which expresses the feelings of every combat Soldier who has ever served.

Soldier Poem (Captain George Skypeck)

Foreword

The imaginative battlefield concepts, ideas and hard-nosed candid analyses in this superbly researched book should be required reading and study by leaders and staff officers who are reorganizing America's Army —— and by Capitol Hill elected officials involved in appropriating money to make it possible. It will save them a lot of time and a lot of money during this critical, watershed transformation period for the Army. The authors, by their monumental work effort and systematized, orderly, readable product have done a great service for the Army and for the above researchers and decision-makers. In many respects, what they have produced in the *Air*-Mech-Strike 3-D Maneuver Warfare field can be called a "*Year 2000*" version of what the *Howze* Board did in 1962, when Secretary McNamara directed the Army to re-examine its aviation requirements with a "*bold new look*" at land warfare mobility. On page 19 of "*Air Mobility 1961-1971, Department of the Army*" by Lieutenant General Tolsen (now deceased), McNamara stated; "*I shall be disappointed if the Army's re-examination merely produces logistically oriented recommendations to procure more of the same rather than a plan for employment of fresh and perhaps unorthodox concepts which will give us a significant increase in mobility*".

The authors cover the history of *Air*-Mech-Strike 3-D Maneuver Warfare; it's use in Vietnam, Panama, and in the Soviet/Russian, British and German Armies. Also thoroughly analyzed is what the book *calls "evolutionary adherence to 2-D heavy mechanized maneuver doctrine*"; and the cost, air/ground vehicular survivability, digital, technological, deployability components of a 3-D capability in various parts of the world. Nothing is overlooked. The contents are presented in a logical manner —well-organized, intelligent, coherent, rational and reasonable. Specific actions are recommended.

As I read through the book, my mind went back several times to my 2 1/2 years in the Air Mobility Division of General Gavin's and later General Trudeau's Office of Research and Development in the late '50s. I was a Major, "*one-man Airborne Branch*", and was involved in the exciting, fast-moving birth of the family of Army Aircraft; Army Air Mobility; new Paratroop and heavy-drop Air Force Aircraft; and new equipment for Army parachutists. My fellow "*action officers*" in ODCSOPS were doing the "*grunt work*" on tactical and strategic concepts. Now 43 years later, I sense a similar dynamic situation in response to the vision and directives of General Shinseki.

By 1964, I was an Infantry Battalion Commander in the 11th Air Assault Division; testing new Air-mobile/Air Assault concepts. From my experience in that duty at Fort Benning for 14 months and in the 1st Cavalry Division (Air-Mobile) in Vietnam as a Battalion and Brigade Commander, the following thoughts;

1) First and foremost, the truly successful and most effective Air Mobile/Air Assault maneuver unit commanders had to be aggressive, self-confident men who thought in 3 Dimensions—in terms of landing zones, pick-up zones; rapid, risky air movements; in terms of empty choppers making diversionary landings to set up the enemy for the kill; close-in and distant battlefield helicopter troop movements to the flanks and rear of the enemy to gain quick tactical advantages. They had to be men with fine-tuned battlefield instincts; men who could **ACT** without second-guessing. Having read this book, *Air-Mech-Strike* 3-D maneuver unit Commanders must be men of that ilk and breed or that type of maneuver warfare will not "*get off the ground*".

Unfortunately as my time in Vietnam went on, I saw several replacement infantry Battalion/Brigade Commanders come into the 1st Cav. who were not capable of the above—who looked on the helicopter simply as an "*air-mobile truck*" to move troops from Point A to Point B.

2) For more battlefield effectiveness, the Air-Mobile Division in Vietnam sorely needed an Airmobile Armored Fighting Vehicle capable of quick tactical moves by air in the battle area. I recall working on development of an air-droppable/*Chinook* transportable Armored Recon Airborne Assault Vehicle (AR/AAV) in OCRD in the late '50s—which turned into an overweight, ground-bound *Sheridan* Light Tank—no longer helicopter transportable. I could surely have used some of the aborted lighter weight "AR/AAVs" in the LZ *X-Ray* Battle in the *Ia Drang*, November 1965.

Finally, I fully agree with the authors of this important, visionary book that the Army should field an *Air*-Mech-Strike 3-Dimensional maneuver warfare capability NOW with existing equipment and technology. I hope the Army gets it in the field with carefully selected Commanders who can think quickly and act quickly in 3 Dimensions. Work it in differing field and tactical conditions and improve it with experience. Do not waste time and money striving for the "*perfect solution*". Use what is now available and move forward.

— Harold G. Moore
Lieutenant General, USA-Ret.
Crested Butte, Colorado
March 21, 2000

Chapter 1

Introduction

"Superior mobility must be achieved if we are to surprise our opponent, select the terrain on which we are to fight, and gain the initiative. There is no alternative. If we are slow in movement, awkward in maneuver, clumsy in deployment — in a word, not mobile — we can expect to be forestalled, enveloped, or constrained to launch costly frontal attacks against an enemy advantageously posted."
—-*Infantry in Battle*; page 106, The Infantry Journal Inc; Washington D.C., 1939

In the late 50s and 60s, a few thoughtful and far-sighted U.S. Army officers conceptualized the integration of Army Aviation into battlefield maneuver. These ideas were rooted in WWII Airborne concepts and driven by advances in helicopter development during and after the Korean War. These officers pushed the creation of combined-arms teams of infantry, artillery and aviation for shock and maneuver. Lieutenant General Gordon B. Rogers chaired the first board called the *"Aircraft Requirements Review Board"*, or the *"Rogers Board"* in 1960. This board focused on the development, procurement and personnel planning for such a capability. It recommended a second board for an in depth study to explore the concept of Air Fighting Units. In 1962, another *ad hoc* task force was put together to re-examine the requirements and the role of Army Aviation. It was known as the *"U.S. Tactical Mobility Requirement Board"*, or as most remember it, the *"Howze Board"*, named after the board's president Lieutenant General Hamilton H. Howze. The board investigated, tested, and evaluated organization and operational concepts of Air-Mobility. It recommended the creation of an Air Assault Division, the 11th Air Assault, to test the concept. This unit deployed to Vietnam in 1965 reorganized as the 1st Cavalry Division[1].
 The Howze Board and the Vietnam War brought about the

beginning of today's modern Army Aviation. However, within two decades, the Army had swung back to heavy mechanization, and General Howze, now retired, called for the necessity of another board to revisit the Air-Mobile capability. General Howze called the utility of light aircraft in Field Manual 100-5 *Operations*, a "*patch-job*" and that "*quite obviously Army Aviation is not contemplated in the manual as a basic tool, as a strong, added, available and sometimes decisive capability, as it emphatically should be*". General Howze argued that the Army's current dependence on armor alone did not provide the required flexibility or mobility[2].

Since the Korean War, our Army has stagnated in Airborne and Air-Assault modernization by not developing *Air*-Mechanized capabilities. Aircraft and vehicles should have been designed, or modified, to move reconnaissance and combat vehicles in and around battlefields by ground and air. In essence, a mobile "*phalanx*", that provided mobility, firepower and protection. Aircraft have been modernized, giving the Armed Forces of the United States the most advanced, fixed-wing and helicopter transport fleet in the world, but with the exception of attack aviation, our Airborne/Air Assault ground force capabilities have changed little. Once parachuted or Air Assaulted onto an objective, or off-set from an objective due to the threat, our ground forces remain foot-mobile and lack ground mobility. By not pursuing the advantage and flexibility 3-Dimensional warfare (above or below ground movement), our operations remain restricted to 2-Dimensional warfare. Currently our limited, 3-Dimensional capability remains almost entirely foot-mobile once employed.

The concept of *Air*-Mech-Strike, the employment of mechanized forces by Airborne and Air-Assault means, is not new. It is an old, yet revolutionary mobility concept, originating just before WWII by theorists in Russia then advanced by combat leaders, such as General James M. Gavin. The U.S. Army initially developed 3-Dimensional warfare for our Airborne forces with a limited capability to airdrop artillery, troop carriers and light tanks but this is now practically non-existent. In contrast, the Russian Army has had an *Air*-Mechanized capability on and off for the last 60 years. The British Army has had one since the late 1970s, and the Germans implemented a very flexible *Air*-Mech force beginning in 1989. Our NATO allies used their *Air*-Mech capability moving into Kosovo in June 1999 as our forces moved into the area on one road via single column. The advantages and flexibility of maneuver that an *Air*-Mech Airborne and Air Assault capable force provide the modern day Commander, makes one wonder why the U.S. Army has not optimized 3-Dimensional warfare. *Air*-Mech-Strike 3-Dimensional warfare provides our Commanders maneuver options and takes advantage of the indirect approach to maneuver the force and win battles.

Joint Pub 3-0 *Joint Operations* emphasizes the necessity for forces that can operate beyond the enemy's ability to react; can overwhelm the enemy throughout the battle area from multiple dimensions in time and space; has operational reach and approach; and when operating on "*exterior lines*" has a mobile force that can converge on the enemy.

Throughout the ages, effective results in war have rarely been obtained unless the approach has had such indirectness to ensure the opponents unreadiness to meet it[3]. Alexander the Great's advance was prepared by a grand strategy of indirect approach, which severely shook the Persian Empire. His success on the battlefield was due to the superior quality of his forces employing both a heavy and a light "*phalanx*" and use of the tactical, indirect approach[4]. Mongol strategy was based on mobility, speed[5] and concentration of forces, both heavy and light cavalry, and it was this trinity of military tactics that led them to tremendous victories time after time. Mongol protection rested in speed and maneuverability throughout the depth of the battlefield. *Air*-Mech-Strike enhances the 2-Dimensional maneuver prowess displayed by Alexander, Genghis Khan, Guderian, Rommel, and Patton with a 3-Dimensional capability. It not only provides ground and air mobility to the war-fighter, it provides the overmatch speed to move to situations of advantage.

Rapidity is the essence of war; it allows you to take advantage of the enemy, to make your way by unexpected routes and attack unexpected spots[6]. Maneuver is a matter of momentum, timing and position. *Air*-Mech-Strike provides that

British CH-47s "*Air*-Meching" vehicles into Kosovo (U.S. Army)

rapidity, that maneuver, both in the air and on the ground—to gain positional advantage relative to an enemy's center of gravity and still retain freedom of action. It allows the force to move without restriction, around obstacles and difficulties rather than through them[7].

The Chief of Staff of the Army (CSA) General Eric K. Shinseki has stated in his vision that our "*heavy forces must be more strategically deployable and more agile with a smaller logistical footprint, and light forces must be more lethal, survivable and tactically mobile*". To achieve this requirement for air-deployed-forces, we propose the use of mechanized-vehicles or an "*Air*-Mech" paradigm. It will require innovative thinking about structure, modernization efforts and spending as the CSA has directed.

"*Joint Vision 2010*" requires that the Army be a strategic responsive force, dominant across the full spectrum of operations. The *Air*-Mech-Strike model provides a responsive force that has the organization and capabilities to operate both in combat and stability and support operations.

Power projection depends on mobility. Speed is essential when deploying our force to the theater of operations. Factors affecting speed are distance and size/weight of the deploying force. A Division's deployable weight averages 60-70% equipment with the remaining 30-40% being warfighting consumables.

Air-Mech-Strike provides optimum strategic deployability into an operational theater regardless of austere conditions and the absence of debarkation ports and airfields. *Air*-Mech-Strike leverages commercial wide-body aircraft (Propositioned Commercial Air) freeing up critical strategic military lift. It allows a combat force to deploy through a Theater Support Base (TSB), Intermediate Staging Base (ISB), Forward Operating Base (FOB) or directly into the area of operations. Realizing that strategic responsiveness is not enough, *Air*-Mech-Strike provides an adaptive, multi-functional force emphasizing the "third dimension" of intra-theater fixed and rotary-wing aviation, allowing the capability to dominate the situation upon arrival through mobility, lethality and survivability. This capability provides early, forced-entry forces, the synergy of rapid maneuver and interdiction on the battlefield with protection.

Strategic analysis of the future threats, and ongoing and past conflicts, lay out a significant range of operational requirements for our Army as we move into the 21st century. If the Army remains relatively the same size, receives approximately the same budget, and continues to maintain the operational tempo of today, it must be very flexible and multi-functional to carry on with the high standards of success expected by the American people.

"*Air*-Mech-Strike 3-Dimensional *Phalanx*" achieves the CSA's vision and fulfills the requirements of *Joint Vision 2010* through a recommended interim and objective force concept tailored for all types of terrain and environments against a myriad of threats.

Air-Mech-Strike provides a flexible, land combat force with the capability of air, mechanized and dismounted maneuver to achieve decisive action through positional advantage regardless of open or restricted terrain. Commanders can increase the tempo of operations with *Air*-Mech-Strike forces to expand the battlefield in space, time and the echelon of forces. The non-linearity and simultaneity provided by this capability, poses multiple dilemmas to the enemy as our forces exploit the effects of fire with maneuver. *Air*-Mech-Strike provides the mobility to find, fix and destroy enemy units faster than they can respond to our synergistic, massing effects.

Battlespace dominance is enhanced with the robust reconnaissance *Air*-Mech-Strike provides the field Commander. This agile air and ground reconnaissance capability bolsters the Commander's understanding of the enemy and terrain. This reconnaissance facilitates initiative allowing the Commander to have forces in the right place, at the right time, with the right mix. Reconnaissance is a precursor to fire and maneuver — the *Air*-Mech-Strike Reconnaissance Surveillance Target Acquisition (RSTA) capability provides both the human and technological capabilities required. Some military experts say movement is the first battle to be won, others say its reconnaissance; either way, *Air*-Mech-Strike provides both to our Commanders.

The Commander has the reach to concentrate firepower and employ his Soldiers, minimizing their foot dependence, unless required in that mode for restricted terrain. The challenges of the Soldier's load are solved with reasoned vehicle support. Soldiers can be moved quickly on the battlefield keeping them fresh and relatively protected during movement. As J.F.C. Fuller said: "*A Soldier cannot be a fighter and a pack animal at one and the same time, any more than a field piece can be a gun and a supply vehicle combined*". The force mix of the *Air*-Mech-Strike unit has the capability to "*break*" enemy armored formations, *fix*, and then *destroy* these formations with combined-arms assets. *Air*-Mech-Strike enhances our Airborne, Air Assault and light forces with sustained combat capability, as well as our heavy infantry with multi-dimensional maneuver.

Lessons learned from the Kosovo air campaign proved again that to destroy an enemy's army you must put "*boots on the ground*". When the enemy dispersed and went to ground, to negate the effects of our technology and precision guided munition advantage, ground maneuver would have been the only means to force him out to be destroyed. There was a requirement to dislocate the enemy from his chosen positions either by forcing him to move from those positions or by rendering his positions useless and irrelevant to the fight.

Getting bogged down with heavy forces trying to fight through extensive, obstacle-laden defiles, or conducting Airborne and Air Assault operations that put foot Soldiers against enemy armor, would have been risky business with the probability of delayed linkups. An *Air*-Mech-Strike force would have provided the means to rapidly conclude the operation successfully. Long-range fires would have been used to paralyze the enemy allowing the *Air*-Mech-Strike force to be inserted by Airborne or heliborne means to cut Lines of Communication (LOC), hit soft targets, and influence key terrain in the enemy's rear. *Air*-Mech-Strike forces would then find the enemy and bring in observed fires on their positions. The positional advantage of our elements would force the enemy to move out of hidden positions located in restricted urban and forested terrain; compelling him to mass, and become exposed to our joint fires and close combat. This *Air*-Mech-Strike 3-D maneuver would be synchronized with a major ground 2-D attack putting the enemy in a dilemma.

During the Gulf War, the Iraqi army used the same technique. In the desert terrain, they knew they could not deal with an overmatched armored fight and the air power of our forces. They went to ground to wait out our firepower-attrition warfare. Not until ground forces attacked were they forced to move from their cover. When they did so they died. They used the same technique in the employment of their *Scud* missiles, using excellent camouflage, and the use of decoys. Our Special Operations Force (SOF) units had to be inserted hundreds of kilometers deep using desert vehicles, internally loaded in medium helicopters; to find, confirm and destroy *Scuds* by directing USAF strikes.

As S.L.A. Marshall highlights in "*Men against Fire*" the essential truth; "*how much fire can be brought to bear is the basic problem in all tactics, and that movement is the means of increasing the efficiency of one's fires.*" *Air*-Mech-Strike provides that "*movement*" to gain a positional advantage over the enemy to simultaneously bring to bear the full effect of all fires; both observed and organic weapons. The effects of these fires produced from the rapid arrival of mounted and air-delivered *Air*-Mech-Strike forces has a tremendous psychological effect on the enemy's C2 and soldier morale. Combined with other engagements (Close Air Support, deep attacks, heavy-force ground attacks and information warfare), *Air*-Mech-Strike dislocates the enemy and places him at such a disadvantage that he must surrender or be destroyed. Thus, the advantage of *Air*-Mech-Strike is eliminating the enemy's capability to fight through destruction and by influencing the imponderables of morale, discipline, leadership and organizational continuity. This puts the U.S. Forces inside the enemy's decision-making cycle.

The *Air*-Mech-Strike Commander has the flexibility, through maneuver options, to preserve freedom of action in all types of terrain and over obstacles either with three-dimensional assets, or if a water obstacle is encountered, swimming the vehicles themselves. When weather restricts flying, the force can continue to maneuver to the vicinity of the objective with armored vehicles. The light overall weight of *Air*-Mech-Strike vehicles allows for continuous movement in countries with very poor road and bridge infrastructure as is found in the Balkans. *Air*-Mech-Strike promotes access to restricted terrain.

Today, information operations has expanded the operational area. This condition is well suited for an *Air*-Mech-Strike organization, which has the capability to exploit this advantage with three-dimensional maneuver throughout the depth of the battlefield. These forces can and should also take advantage of our Army's expanded situational awareness. Expected tactics are normally used to confront the opponent, but it is the power created by unexpected tactics, that is, innovative use of people and information that makes victory certain[8]. *Air*-Mech-Strike provides those unexpected tactics. *Air*-Mech-Strike gives our commanders a robust heavy or light, fixing force and a mobile, maneuver force.

We must continue to have a first-rate Army, capable of operating across the spectrum of conflict. This full-spectrum Army organized around a common Divisional design, with an *Air*-Mech-Strike capability, is the answer. The following chapters lay out that flexible organization and capability.

The success of Mongol tactics resembled those of the hunter, who uses speed, finesse and deception to herd his prey where he wills, then kill it with as little risk to himself as possible[9]. *Air*-Mech-Strike allows our Commander to be the hunter—not the prey.

---

[1] Lieutenant Colonel Arthur France, *Combined Arms in Battle since 1939*, (Fort Leavenworth, Kansas: U.S. Army Command and General Staff College Press, 1992), pp. 11-12

[2] *The Howze Board*, (Arlington, Virginia: *Army* magazine, April 2000), p. 15

[3] B.H. Liddell-Hart, *Strategy*, (New York: Meridian books, 1991), p.5

[4] Ibid. p. 144

[5] David S. Woolman, *Primitive Warriors or not, the hardy Mongols of the 12th and 13th centuries used the most advanced of tactics* (*Military History* Magazine, October 1995) pp.12-16

[6] Sun Tzu, Introduction by Samuel B. Giffith; *The Art of War* (New York: Oxford University Press, 1984) p.23

[7] USSOCOM Publication 1, *Special Operations in Peace and War*, (MacDill Air Force Base, Florida: U.S. Special Operations Command, January 1996) pp.5-4

[8] Sun Tzu, Introduction by Samuel B. Giffith; *The Art of War* (New York: Oxford University Press, 1984) p. 35

[9] Erik Hildinger , *Mongol Invasion of Europe* (*Military History* Magazine, June 97) pp. 39-42

Chapter 2

Purpose

"Nothing is more difficult than the art of maneuvering for advantageous positions... In all fighting, the direct method may be used for joining the battle, but indirect methods will be needed in order to secure victory."
—Sun Tzu

The purpose of this book is to provide a means to ensure victory throughout the spectrum of operations with enhanced maneuver; to increase the Army's versatility through Division reorganization; to emphasize the development of robust reconnaissance at all echelons; and to take advantage of our Army's existing tracked vehicle inventory and large rotary-wing fleet.

This is accomplished with an *Air*-Mech-Strike 3-Dimensional warfare capability. The proposal described in this book is cost-effective for the Department Of Defense (DOD); is motivating to our Soldiers; emphasizes offensive action; provides a means to win against all enemies—in all-terrain, and is supported with lessons learned throughout military history. Most importantly it gives us the ability to move.

Major General JFC Fuller's analysis of Armies in conflict defines that battles are fought in accordance with three defined tactical functions: to move; to hit; and to guard. These functions are supplemented by others: to find; to hold; to destroy/pursue. An Army's ability to move underlies these functions. The weapons deployed fulfill the object of the functions. Secure and rapid movement is the aim of the maneuver battle.[1] *Air*-Mech-Strike allows us to fight battles on the move—faster than the enemy. This *Air*-Mech-Strike proposal supports the U.S. Army Chief of Staff (CSA), General Eric K. Shinseki's 1999 transformation vision statement:

" The Army will develop the capability to put combat

forces in Brigade Combat Teams anywhere in the world 96 hours after liftoff for both stability and support operations and for warfighting. We will build that capability into a momentum that generates a warfighting Division on the ground in 120 hours and five Divisions in 30 days. Operationally, we must have the capability to position forces to create advantage in the theater and on the dispersed battlefield of the future. Our information superiority can help create that advantage, but our forces must be rapidly deployable in three dimensions across the theater, providing an adversary a more complex targeting challenge and enabling us to accomplish the full range of operations rapidly and decisively"[2]

This book proposes a solution to the challenges in the Army's transformation vision, combining 2-D/3-Dimensional maneuver capabilities that are *"persuasive in peace and dominant in war"*. We outline how the U.S. Army can build today, at extremely low cost, an OPTIMUM full spectrum-of-war force using existing vehicles, and a modest buy of off-the-shelf platforms, leveraging current *"technology insertions"*. The *Air*-Mech-Strike (AMS) force structure could be built today by industry and the Army, with the first unit fielded within 12 months.

The Army not only needs medium-weight forces, it needs 3-Dimensional *Air*-Mech maneuver capability in all its Divisions. With the enemy often predicting our 2-Dimensional heavy attack routes; our extensive requirement for lodgment and buildup; the probability of operating in urbanized and other restricted terrain; the advent of signature-less, top-attack, stand-off munitions; and media pressure time factors for deployment, the U.S. Army needs multi-dimensional forces to prevail. We need *Air*-Mech-Strike Divisional, Brigade and Battalion formations to negate these challenges.

Why *Air*-Mech-Strike 3-Dimensional Phalanx?

"Since the beginning of time man has sought to gain a mobility advantage over his opponents in war. The Greeks developed

the phalanx, which enable them to mass their arms and men in such a way that they moved as an integrated unit of power"
— Lieutenant General James M. Gavin, page vii, *Air Assault: The development of Air-mobile Warfare*, by John R. Galvin (USA, ret.), Hawthorne Books, New York, NY, 1969

Air-Mech-Strike provides the capability to maneuver ground and air—over any terrain, mechanized forces with sufficient combined-arms to operate throughout the spectrum of conflict. This requires fixed and rotary winged aircraft that can move by strategic deployment and tactical employment—combat, combat support and reconnaissance vehicles. The ground force moved, or supported, is called the *"Phalanx"*.

What is a *"Phalanx"*?

In ancient armies, Soldiers had specially-made shields which could be joined together as the men marched in battle formation to create in essence, a human armored vehicle or "tank". Called a *"Phalanx"* this group armor would protect against missiles in the form of arrows, javelins, spears and rocks. As this formation closed with the enemy its shock action could crush men loosely arranged as the infantry under the *Phalanx* could use their spears or swords sticking out from their shields. Soldiers in the center could place their shields above their heads to protect against top-attack javelins and arrows, giving all-around protection. While the *Phalanx* was difficult to use in uneven terrain, the practical military commander didn't break it and throw it away. He modified and improved it. When Philip of Macedon, improved the *Phalanx*, his son, Alexander the Great—used it to conquer the known world. He transformed his Army into one of the most effective land combat Armies in world history. The Romans, improved upon the *Phalanx* by making it smaller, and employed it as one of many possible formations as the situation required.[3]

 The *Phalanx* is an excellent description of the infantry-centric force the U.S. Army needs to build, because for it to work best, all of the parts or "shields" have to be designed to fit together. We want to improve upon the *Phalanx* like the Romans did to the Greek formations, giving our modern-day *Phalanx* flexible, combined-arms teams to execute *Air*-Mech-Strike 3-D operations. It represents a force that can move quickly and under armor protection that can get to the scene of the fight. It can stay together to be "armored", or separate into smaller "infantry" units to fight or keep the peace. It has the flexibility to leave its armor due to restricted terrain, yet can quickly return to its armor for protection. This 21st Century "*Phalanx*" will be a force that has parts that can fly over the heavy *Phalanx*, moving in concert against our enemies. Our *Air*-Mech-Strike 3-Dimensional *Phalanx* is a means to maneuver with the main, heavy 2-D force *Phalanx* emphasizing non-linear operations in restricted terrain.

An *Air*-Mech-Strike maneuver confuses the enemy since he does not know which point to defend. To hit the enemy's army we want at least two forces just as a boxer wants two fists, as all attacks should, when possible, be dual in nature, since it is more difficult to watch two fists than one.[4] Because speed is a major factor, our forces can take advantage of a situation before the enemy can react. *Air*-Mech-Strike can create a "*third flank*" with our Airborne/Air Assault armor. This maneuver emphasizes speed and movement to present an opponent with a rapidly developing and quickly changing situation.[5] Improved situational awareness provided to *Air*-Mech-Strike units supports the distribution of advancing forces while maintaining the unity of combined action. The cumulative effect of small successes or even the mere threat, at a number of points, may be greater than the effect of complete success at one point.[6] The take-down of Panama in 1989 is prime example. Adoption of this type of maneuver warfare means a fundamental change in how our Army is used to fighting. *Air*-Mech-Strike maneuver requires flexibility and decentralization. It is the combination of offensive and protective power with a capacity to move — much like a tank's fighting strength — but with an air capability.

Air-Mech-Strike also fulfills the operational requirements of Joint Publication 3-0:

—Integrate and synchronize operations in a manner that applies force from different dimensions to shock, disrupt and defeat opponents (**Synergy**)

—Bring force to bear on the opponent's entire structure in a near simultaneous manner to overwhelm and cripple enemy capabilities and the enemy's will to resist (**Simultaneity and Depth**).

—Mass effects against the enemy's sources of power, gain positional advantage, concentrate at decisive points to achieve surprise, psychological shock and physical momentum (**Centers of Gravity**).

—Operate from overseas, sea-based or continental basing; reach operational area regardless of geography; conduct forced entry operation with speed, surprise and the ability to fight on arrival (**Operational reach and approach**)

—Exploit the effects of massed and/or precision firepower (**fire and maneuver**)

—Synchronize interdiction and maneuver (**complementary operations**)

—Conduct multi-national operations (**capability to work with NATO/Allied *Air*-Mech forces**, ie. British, German, Russian)

—Establish and **operate for a Joint Force Land Component Commander** —JFLCC (*Air*-Mech-Strike Forces strengthen a Marine Air-Ground Task Force, MAGTF, working littorals by operating farther inland, expanding the depth of the battlefield)

—Conduct sustained operations, dimensional superiority (**Mix offense/defense, mix linear/non-linear, synchronize maneuver/ interdiction**).

—Adapt to advances in technology that continue to increase tempo, lethality, and depth in warfare (**Revolution in Military Affairs-RMA**)

—Operate from exterior lines with stronger, more mobile forces (**3-D Maneuver**)

—At the strategic level, have a capability to threaten invasion; have appropriate mix of capabilities for a joint force; have flexible combat power across all dimensions; have the capability to take advantage of unexpected opportunities to exploit (***Air*-Mech-Strike organization**)

—At the operational level—set the terms for battle by time and location, be able to cut Lines Of Communications; seal off retreat; dislocate enemy command and control; shape the operational area prior to engagement (*Air*-**Mech-Strike capability**)

Reconnaissance

"No activity is more closely tied to success than effectively gathering and disseminating intelligence".
—Sun Tzu

The reconnaissance organization for the *Air*-Mech-Strike force and for the recommended Brigade and Division organizations is very robust. Our Army has lacked sufficient reconnaissance for decades; as demonstrated time and again during operations and during Combat Training Center (CTC) rotations. This deficiency must be fixed. Reconnaissance shapes the battle for victory—consequently a significant change in our Army's Reconnaissance Surveillance Target Acquisition (RSTA) structure is recommended. USSOCOM Publication 1, stresses the requirement to *"find enemy weaknesses and vulnerability; avoid enemy strengths and reduce battlefield uncertainty"*. Our reconnaissance organization recommendation will do just this. Relevant Intelligence and Information (RII) is the supporting base for Information Operations (Civil Affairs, Psychological Operations, Deception, Direct Action and Public Affairs). Our RSTA organization is flexible and adaptable to requirements imposed by objectives and terrain; focused on gathering, sharing and using intelligence; and taking advantage of the critically of time. This is essential since today's battles are information battles where 70% of the value of information is gained from timeliness and information determines perception and opinion, which affects national will.

Vehicles

The Army is transforming itself partially to medium-weight armored vehicles. This book proposes that those combat vehicles are tracked and that they weigh under 11-tons for terrain agility and air movement. Tracked vehicles are superior in off-road movement, provide better protection, have a supportability advantage and carry more infantry than equal-sized wheeled vehicles. We propose the upgrade of the existing inventory of M113s (named the *Gavin* Fighting Vehicle)[7]. There are 25,000 M113s in our inventory and the upgrade cost to convert these vehicles from M113A2 to A3 is only $150-310,000 per vehicle depending on how the conversion is done. If one piece rubber "band track" is purchased, along with a composite ramp, the M113A3 has a CH-47F lift capability. The M113, in any variation is C-130 transportable. Also proposed, in this book, is the purchase of the M8 Armored Gun System (AGS) with a shoot-on-the-move 105mm-gun capability. The M8 is C-130 airdrop capable now. The Army has already spent $500 million for production preparation and millions for an entire family of 105mm ammunition stockpiled and awaiting use.

Two vehicles are recommended for reconnaissance. The available off-the-shelf, German *Wiesel* (we call it the "*Ridgway* Fighting Vehicle"), a tracked four-ton helicopter-transportable armored vehicle and an All Terrain Vehicle (ATV) for stealth. These reconnaissance vehicles are deployable by either commercial aircraft or military strategic air.

Heavy M1/M2 fighting vehicles remain a critical part to our recommended Divisional structure. This heavy capability is balanced with the lighter *Air*-Mech-Strike vehicles providing flexibility in task organization for any type of operation required, any terrain and optimum deployability.

Interim and objective force organization models are recommended for the Army that provide an immediate capability with existing Army vehicles and aircraft, and a future capability as technology requirements produce an optimum *Air*-Mech force.

We propose a change in every Division of the Army, not only for a few Brigades. This is essential for flexibility, for an informative-based force, for regional orientation, for global reach, and the reality that we have a small Army with many commitments. Sun Tzu concluded; "*He will conquer who has learned the artifice of deviation. Such is the act of maneuvering.*"[8] *Air*-Mech-Strike is the means by which today's U.S. Army can boldly maneuver with maximum flexibility of action.

[1] Brian Haldon Reid, *MG J.F.C. Fuller and the problem of military movement*, (Fort Knox, Kentucky: U.S. Army *Armor* magazine, July-August 1991); pp. 26-31

[2] Chief of Staff, General Eric K. Shinseki, Speech to AUSA, October 1999 (Fort Bliss, Texas, U.S. Army Air Defense Artillery magazine online) *http://147.71.210/adamag/Oct99/CofS.htm*

[3] Compton's online Encyclopaedia: *http://www.optonline.com/comptons/ceo/00286_A.html*

Philip II of Macedon did a major Army reorganization and created **one of the most effective land-based fighting units in history**. He changed the structure of the *phalanx* by making it 16 lines deep instead of eight. A single division of *hoplites* numbered 4,096 —16 lines of 256 Soldiers each. Preceding the hoplites into battle were four lines of 256 psiloi, light infantry. Behind the *phalanx* division were eight lines of 256 *peltasts*, also light infantry. Cavalry units covered the flanks. Including cavalry, a full division consisted of 8,192 men. Attached to the Army were a medical corps and a corps of engineers. This Army was inherited by Philip's son, Alexander the Great, and used to conquer a great portion of the Mediterranean world (see Alexander the Great).

Gary Brueggeman, The Roman Army:
http://www.geocities.com/Athens/Oracle/6622/legions.html

By the time of Caesar it is clear that the *cohort* had become the fundamental tactical unit. He describes battle formations and troop movements in terms of *cohorts* and *legions* appear more as administrative units than battlefield tactical units. Caesar sends *cohorts* and groups of *cohorts*, not *legions*, on flanking maneuvers. Since the number of men in a *cohort* could vary considerably the formation must have been scaleable in some consistent manner so that its fighting efficiency was not destroyed by the reduction in manpower. The typical formation, described above, uses a spacing of 0.91m (3') for the files and 1.22m (4') for the ranks, allowing for the *centurion*, then, the cohort would have a front of 15.85m (52'). If the three cohorts fought side by side, as is always depicted, then their combined front would have extended to 48.16m (158'). The *cohort* would, in essence, be a *miniature phalanx*, 10 ranks deep and 58 files wide.

[4] J.F.C.Fuller, *On Future Warfare* (London: originally printed in 1928, reprint 1980) p. 80

[5] Wayne Downing, General, *Firepower, Attrition, Maneuver— U.S. Army operations Doctrine: A Challenge for the 1980s and beyond*, (Fort Leavenworth, Kansas: U.S. Army *Military Review* magazine, January-February 1997)
http://www-cgsc.army.mil/milrev/english/janfeb97/downing.htm

[6] B.H.Lidell-Hart, *Strategy*, (New York: Meridian, 1991), p.333

[7] James M. Gavin, *Airborne Warfare*, (New York, Infantry Press, 1947), pp. 174-175

[8] Sun Tzu, Introduction by Samuel B. Griffith; *The Art of War* (Oxford University Press, 1984); pp. 10-51

4-Ton RFV Strategic 747 Air Mobility Frees C-17 airlift for Heavy Combat Modules

4-Ton RFV
Air-Mech Model

20 X C-17s

M-BDE TF 280 RFVs
25 X B-747s

Heavy TF
MLRS
Lift Aviation
Attack Aviation
Patriot
Engineer

4-ton Model for Interim U.S. AMS Forces frees C-5s/C-17s for outsized forces (Major Charles Jarnot)

Chapter 3

Understanding U.S. Army 3-D Maneuver Warfare History

The Chief of Staff of the U.S. Army has declared: *"The world situation demands an Army that is strategically responsive. The Army's core competency remains fighting and winning our Nation's wars; however, the Army must also be capable of operating throughout the range of conflict — to include low-intensity operations and countering asymmetric threats. It must, therefore, be more versatile, agile, lethal, and survivable. It must be able to provide early-entry forces that can operate jointly, without access to fixed forward bases, and still have the power to slug-it-out and win campaigns decisively. At this point in our march through history, our heavy forces are too heavy and our light forces lack staying power. Heavy forces must be more strategically deployable and more agile with a smaller logistical footprint, and light forces must be more lethal, survivable, and tactically mobile. Achieving this paradigm will require **innovative thinking about structure, modernization efforts, and spending.**"*[1]

The U.S. Air Force provides the U.S. Army Airborne forces with excellent strategic mobility. The U.S. Army's Aviation Branch has provided its Air Assault forces with the ability to engage in decisive operational and tactical maneuver. However this mobility and maneuverability ends when the Airborne or Air Assault force lands on the ground. Then, the Paratroopers and Air Assault forces become marching foot Soldiers who must carry their own weapons plus essential logistics. Accompanying supporting weapons, such as tanks, armored personnel carriers and heavy artillery, are unavailable or at best in short supply, so the inserted Airborne or Air Assault force lacks extensive combat power and is often dependent on the ability of supporting aircraft to stay overhead or react. Since these forces lack rapid, long-range ground transport, they must land on or close to their objective. This limits the types of objectives which they can take to those which are lightly-defended or undefended. Landing on or near the objective creates its own set of problems. The landing force must rapidly assemble, organize and deploy before the enemy can react. This means that force has to achieve tactical and operational surprise or face a rugged fight. This lack of ground transport limits American Airborne and Air Assault forces in the types of missions they can undertake and the types of forces they can successfully defeat.

The need to combine ground and air mechanized elements to create full 3-Dimensional maneuver has been understood since the beginnings of infantry delivered by aircraft and the armored tank. However combining these arms has not been easy, German Army General Von Toma wrote in 1945; "*The lack of integration and cooperation between our Infantry and Armor led to our eventual downfall*"

The Drift Towards Heavy 2-D Maneuver

"Ironically it was the Germans, operating under heavy restrictions imposed by the armistice of 1918, who proved to be the strongest Air-mobile power by the beginning of the Second World War."
—General John R. Galvin (Retired), former NATO Commander, page 312, ***Air Assault****: The development of Airmobile Warfare*

The beginning of modern mechanized maneuver begins in World War I. The armored tracked tank was created to break the deadlock of the trenches in WWI by men like Churchill, Swinton and Fuller. The battle of Cambrai in 1917 was the centerpiece of a primitive infantry/artillery/aircraft combined-arms team that gained 2-D/3-D maneuver to end the war in victory for the Allies. Fixed-wing aircraft could take-off and land from grassy fields and carry large bomb loads or a dozen armed men with parachutes. In fact, had the war not ended, the U.S. Army would have launched the first large-scale Airborne operation. General Galvin writes:

> "The date was October 17, 1918, less than a month before the armistice. General Jack Pershing had led his forces through the highly successful battle of Saint-Mihiel, in which the American Expeditionary Force (AEF) captured fifteen thousand Germans and 250 artillery pieces in one thirty-six hour period. On that morning Pershing, anxious to get off to the front, which was about five miles north of Verdun, was being delayed by a discussion with his aviation officer, the flamboyant Colonel Billy Mitchell.
> Pershing wasn't interested in striking enemy's rear with aircraft and he wanted
>
> Mitchell to gain air superiority to drive off German aircraft bombing and strafing attacks. After that, Mitchell's pilots should seek out and strike the German troop concentrations that threatened his forces....After these missions were accomplished, if Mitchell's pilots had any time left, they should go out and get more information about enemy troop movements along the front. Mitchell quickly agreed to put those orders into effect right away, but he was already four or five steps beyond Pershing's carefully outlined concept of local air superiority and close air support. He was in fact making long-range plans for the use of his air power and mobility. By the Spring of 1919, Mitchell assured Pershing the production of bombers would be in full swing and there would be enough of them available to support the all-out drive into the heart of Germany, beginning with an attack toward Metz. The Germans had moved strong forces into position to protect the city, but there was, Mitchell insisted, a way to take Metz without a costly and slow "battering-ram" approach. For one single massive operation, he would gather in all the available aircraft – sixty bomber squadrons consisting of twelve hundred

Handley-Pages, Capronis, and de Havillands – and for this mission they would drop not bombs, but parachute troops. In Mitchell's plan, each plane would carry ten men equipped with parachutes and two machine guns; these aircraft, taking off from many airfields, could simultaneously drop a whole Division; twelve thousand men, behind the German lines. Imagine, he said, 2,400 machine guns ready to create havoc at a strongpoint in the German rear, while the Allied main attack moved through the crumbling and panicked enemy front lines. To do the job, Mitchell said he wanted the 1st Infantry Division. He would lift the Division into position behind the German lines in one great air armada, with the bombers protected by hundreds of fighter planes flying above, below, and beside the formation. Other pilots, flying at low level all around the position, would disrupt any German counterattacks while the American troops landed, dug in, and prepared themselves. He would keep the strongpoints supplied by air. "It is conceivable that all would not have landed safely, that not every platoon could have been reformed behind those German lines," Mitchell admitted later, "but remember, we should have had a potential strength of 2,400 machine guns. If we could have only got 10 per cent in action against the enemy's rear we should have been successful." Pershing was skeptical that such an attack could be mounted, but he surprised Mitchell by giving his tentative approval to the general concept. He told Mitchell to go ahead with detailed planning, and to be ready to explain just how it would be done and how the necessary resources would be marshaled and used. In the meantime, said Pershing, he would be happy just to have American control of the air over the expeditionary force. After Pershing left for his daily tour of the front, Mitchell rushed back to his headquarters to talk to his new operations officer, Major Lewis Brereton. Mitchell excitedly explained that he had sold Pershing on the idea of an air assault behind the German lines at Metz. Now all that remained was to work out the details and get the airplanes. It would take every bomber and practically every pursuit aircraft in the Allied fleet, including those promised from production by spring, but Mitchell was sure that if he could show a workable plan, he would be given the Aircraft. In the days that followed, Mitchell carried out Pershing's orders to protect the ground troops in day-to-day operations, and at the same time he kept up the steady pressure to get more bombers into the theater. Brereton made a list of the airfields with range of Metz and figured the maximum number of planes that could be supported by each of them... However, before Bereton could become too deeply involved in attempting to hurdle the many obstacles that confronted him, the armistice stopped his work."

General Mitchell was not the first tactician to exploit the possibilities of troop carrying aircraft since the French had carried out small parachute raids behind enemy lines as early as the spring of 1918, dropping two-man demolition teams to destroy communications. However, Mitchell was the first who thought of Air-Meching an entire Army Division by parachute.

After WWI, the Germans vowed to not be outmatched again. The track-laying armored vehicle did not reach its practical potential until WWII when the Germans improved their performance and provided them mobile infantry in half-tracks (*panzergrenadiers*) to exploit gains and provide close-in protection[2]. The fixed-wing airplane and Soldiers delivered by parachutes and airlanding while created just prior to WWI did not reach practical utility until WWII. The German Army was the first to combine both 3-D Airborne and 2-D Armored warfare with their early string of victories in WWII with "*blitzkrieg*" or "lightning war", the Airborne part of blitzkrieg did not employ a small armored fighting vehicle due to transport aircraft limitations. However, the Germans squeezed a primitive *Air*-Mech-Strike package into the 3-engined fixed-wing JU-52 that was decisive in the WWII battle of Crete. German motorcycle half-tracks and 75mm recoilless rifles were airlanded from JU-52 aircraft. Although many were hit by enemy fire and damaged or destroyed by the waiting, numerically superior Allies, the Germans were eventually victorious[3].

The **"AIR" element was provided by** JU-52 tri-motor aircraft that could airdrop using the left jump door, the bomb-bay or wing shackles. The JU-52 could also airland the German *fallschirmjaegers* or Mountain infantry and their equipment out the left door and larger items through the right cargo doors[4]. The **"MECH" ground mobility was provided by** *Kettenrad* SDKFZ2 motorcycle half-tracks that were just small enough to be delivered by JU-52s through the right fuselage cargo doors. *Kettenrad* SDKFZ2 motorcycle half-tracks then transported Paratroopers and recoilless rifles across the island. The **"STRIKE" was rendered by the** LG 40 75mm Recoilless Rifle which the Allies had not considered as a heavy fire support asset available to the German Airborne. Noted military historian Ian Hogg concludes that the LG 40 was the decisive ground weapon used at Crete.

Following the heavy casualties on Crete, Hitler de-emphasized Airborne operations not knowing the Allies had broken German codes and had been forewarned about the drop onto Crete[5]. Without a 3-D element in German warfighting from 1941 onwards, it's clear that they were then limited to just 2-Dimensions.

The True Father of *Air*-Mech?

To most Westerners, Brigadier General Richard Simpkin and General F. M von Senger und Etterlin are considered as the forces behind the western *Air*-Mech concept.[6] However, Army Commander and later Marshal of the Soviet Union, Mikhail N. Tukhachevskiy was the first father of the *Air*-Mech concept and the Soviet Airborne in the 1930s. Tukhachevskiy was influenced by his experience in the linear fighting of World War I, the fluid non-linear fighting of the Russian civil war and the mobile warfare of the War with Poland. Tukhachevskiy was taken with the potential of Airborne forces in 1928 and began planning for their formation and employment for use in deep battle. From the beginning, Tukhachevskiy realized that Airborne forces needed mobility once they are inserted and planned for the introduction of transport and accompanying combat power for the Paratroopers. Tukhachevskiy proposed that Paratroopers be equipped and organized as Airborne motorized assault forces. Writing in 1931-1932, Tukhachevskiy stated:

> *In our time, air power has grown into a very powerful military factor. During the Imperialist War [World War I], air power supplemented actions by land or naval forces. In the future, air forces can undertake ... independent actions on a much larger scale in coordination with land and sea forces. These independent actions include bombing missions and Airborne landings. Assault troops will land by parachute or be airlanded on suitable areas. Heavy-lift aircraft make it possible to constitute a new type of Airborne motorized assault Division. These aircraft can land Airborne motorized assault forces and maintain combat contact with them. Many people do not believe in the importance and feasibility of this completely new and unusual concept and foresee only sporadic employment of Airborne motorized assault forces. Meanwhile, the British and Americans are rigorously working on the concept and have enjoyed much success. The question remains, will the motherland embrace this new type of war and prepare for it or simply lose the opportunity? If the motherland decides to employ such a force only sporadically, then she stands to lose. If the motherland builds and deploys significant Airborne motorized assault forces, she will be able to seize enemy rail lines on the main approaches and prevent their use by the enemy. She will be able to paralyze the deployment and mobilization of enemy forces. She may overturn the existing operational methods and give the outcome of war a far more decisive character.[7]*

What then is *Air*-Mech? *Air*-Mech is the marriage of aircraft and armored vehicles to combine tactical mobility and combat staying power in a force which can fly deep into hostile territory and conduct decisive maneuver combat.

Tukachevskiy saw the need for *Air*-Mech and tried to build it into the Red Army. When he commanded the Leningrad Military District, he formed an experimental air-landing detachment consisting of a rifle company, sapper platoon, communications platoon and light vehicle platoon plus a heavy bomber aviation squadron and a corps aviation detachment. The unit had 164 men, two 76mm guns, two T-27 tankettes, three armored cars, four grenade launchers, three light machine guns, ten trucks 16 automobiles, four motorcycles and a bicycle. Twelve TB-1 bombers and ten R-5 light aircraft provided the aviation support. Initially, the unit trained and tested the air-landing concept, but did not address airdrop.[8]

In June 1931, Tukachevskiy created an experimental parachute detachment to test airdrop. This new detachment became the parachute element of the Leningrad Military District's Airborne motorized assault force. This experiment in Air-Mech continued under Army Commander I. P. Belov, Tukachevskiy's successor in the Leningrad Military District. Belov echoed Tukachevskiy's proposal to create *shtatny* [table of organization and equipment-TO&E] Airborne Divisions consisting of a motorized Air-landing Brigade, an Aviation Brigade, a parachute detachment and necessary support units.[9] Tukachevskiy's subsequent assignment as Director of Armaments allowed him to develop equipment for *Air*-Mech and mold and shape the concept.[10]

By the end of 1931, the Soviet Union had conducted over 550 Airborne exercises to test the new concept. In 1932, the Red Army issued the draft "*Regulation on the Operational-Tactical Employment of Airborne Motorized Assault Forces*". The regulation stated that Airborne motorized assault detachments were Army operational-tactical units that coordinated closely with ground forces. They would perform wartime diversionary missions such as destroying enemy rail and road bridges, ammunition warehouses, fuel dumps and aircraft on forward airfields. They would support ground offensives by destroying enemy lines of communication, supply depots, command posts and other important rear-area facilities. They would block enemy withdrawal or reinforcement and, during the defense, would attack enemy command posts, disrupt enemy troop movements and capture airfields in the enemy rear.[11]

The Chief of Airborne Services of the Red Army Air staff, I. E. Tatarchenko, supported Tukachevskiy's views. Tatarchenko recommended the use of multiple landing zones in a single operation to lessen vulnerability and confuse the enemy. An Airborne operation would begin with a parachute landing of special reconnaissance, communications and engineer troops who would prepare the landing zones. This advance guard, armed with machine guns, light artillery pieces and anti-tank guns, would secure the landing zones while the main force landed by parachute or by air-landing. Once the main force was on the ground, artillery, vehicles and tanks would be flown into the landing zone. Following this, aircraft would fly in supplies and ammunition. The force would form and begin operations in coordination with those of the main ground effort.[12]

In 1933, the Soviets formed the 3rd Air-landing Brigade in the Leningrad Military District. This TO&E unit had a Battalion-sized parachute detachment, a motorized/mechanized Battalion-sized detachment, an artillery Battalion, and an air group of two squadrons of TB-3 modified bombers and a Squadron of R-5 aircraft. By the beginning of 1934, the Airborne force structure included the Air-landing Brigade, four Airborne motorized assault detachments, 29 separate air-landing battalions and several Company and Platoon-sized elements totaling 10,000 men.[13]

Soviet exercises in the 1930s featured parachute drops and air landings. The 1934 summer maneuvers featured a drop of 1,500 Paratroopers.[14] Tukachevskiy's concept of Airborne motorized assault forces was now a prominent part of the Red Army. However, the Marshal was arrested during the Stalinist purges and was shot in 1937. The Marshal died, but the Red Army Airborne forces lived on. Right up to the beginning of World War II, the Red Army experimented with ways of delivering vehicles with the Paratroopers. In 1941, the Antonov aircraft design bureau built the semi-successful KT-40 bi-plane glider to deliver the T-60 light tank to the drop zone.[15]

The Soviet Airborne forces record during World War II is mixed. *Air*-Mechanization was put on hold. Airborne forces fought the Germans in deep battle on the Moscow approaches during June 1941-January 1942, but most of the original Airborne forces were soon used up fighting the Germans as regular infantry. New units were formed and integrated. By 1944, the Red Army included a separate Airborne Army composed of three corps and nine Divisions.[16] A total of ten Airborne Corps were formed during the war.[17] Most of the Red Army Airborne units fought as infantry, although there were several Airborne drops, including the massive (and disastrous) Kanev cross-Dnieper Airborne operation. Some of the most successful Red Army Airborne operations of World War II were air-landings in the Far East against the Japanese Kwantung Army.

The Quest for 3-D Maneuver Warfare in the U.S. Army

"Thanks to the prodigious efforts of all those involved, and to the courage of the individual troopers, we were able to expand our Airborne forces from what was a platoon lifted by a few aircraft to a complete Airborne Army by 1945. ...we are inclined now to take for granted the Air-Assault capability developed in World War II, but it was not an easy thing to achieve. In the first place, there were many in the 1930's, and even in the early forties, who believed that the next war would be a war of static positions"
—Lieutenant General James M. Gavin, pages vii and viii of General Galvin's, **Air Assault**: *The Development of Airmobile Warfare*

Consider the paths of two American giants, Generals James M. Gavin and George S. Patton. Today, both of these heroes are virtual "patron saints" in the Airborne and Armored communities. Both men appreciated the attributes of the other's style of warfare. To the credit of the Allies, knowing the ULTRA interception had tipped them off to the Crete landing, resulting in heavy German casualties, the U.S. Army believed in the Airborne concept and in conjunction with its Allies built up Airborne forces into a multi-Division "Airborne Army". These 3-D maneuver elements were used to "seize and hold" key mobility corridors and "*force entries*" for amphibious landings, resulting in reduced casualties and great success[18].

In his book, "*War as I knew it*", Patton highlighted his use of Airborne units to seize river crossings and facilitate the advance of 3d Army's armored columns[19]. He stated:

> *The trouble with the Airborne Army is that it is too ponderous in its methods. At the present stage in Airborne development, it is my belief that one Airborne Regiment per Army, available on twelve hours notice, would be more useful than several Airborne Divisions which usually take several weeks to get moving. Three times in our crossing over France, plans were made to use the Airborne Divisions, but we always got to the place they were to drop before they could get ready to drop.*

Likewise, General Gavin was a firm believer in tanks. He used them for fire support during the WWII Njimegan Bridge crossings. He knew first-hand what it was like to fight tanks without supporting arms. At Biazza Ridge on Sicily, he had to use 75mm pack howitzers to stop Tiger tanks[20]. He pushed for and got a glider-delivered 57mm anti-tank gun that could be towed by a jeep. He wanted a disposable recoilless rocket launcher like the *Panzerfaust* that Paratroopers could carry to kill enemy tanks. Most importantly, he wanted tracked armored fighting vehicles for Airborne units.

After the war, Gavin's writings and his work as Army Chief of Research/Development led to the development of the M113 and M551 air-deliverable Armored Fighting Vehicles. Clearly General Gavin was a combined-arms thinker and futurist beyond the "*Jumping Jim*" label applied to him.

The U.S. Army has undergone many major organizational changes since the end of World War II. General James Gavin in his book, "*Airborne Warfare*" in 1947 and "*Cavalry, and I Don't Mean Horses*," in Harper's Magazine, April 1954, have called for air-delivered light tracked Armored Fighting Vehicles (AFVs). This would be the logical development from U.S. Airborne experiences in WWII which were based on a "*seize and hold*" mentality. Operational maneuver from the air; combining the lessons of Airborne and Armored warfare into one, would provide a stronger whole. However, the rush to demilitarize, padded by the false sense of security in the U.S. gave decision-makers no sense of urgency to act on General Gavin's ideas.

The early piston-engined helicopters introduced in the Korean war were limited to small-scale troop transport and medical evacuation. They clearly were not capable of transporting armored vehicles.[21] however; the Airborne did parachute jeeps and towed howitzers along with their assault troops from fixed-wing aircraft. Lieutenant General John J. Tolson noted (in his 1973 paper on airmobility) that the Army's first steps towards airmobility took place in 1952, when it proposed forming twelve helicopter battalions even though there were no troop carrying helicopters or tactics/techniques available to execute the concept. When the Korean War began, the Army owned 668 light aircraft, (mostly artillery spotters), and 57 light helicopters.

In the late 1950s, the Army's Armored Reconnaissance Airborne Assault Vehicle (AR/AAV) program attempted to be an armored vehicle of less than 10-tons. The AR/AAV would do reconnaissance for the main body of Army 2-D units, which had M48/60 Main Battle Tanks and M113 Armored Personnel Carriers. The AR/AAV would be also be a fire support vehicle that could be easily parachute airdropped and CH-47 *Chinook* helicopter transported for Airborne and Air Assault 3-D units.[22] Unfortunately the fielded AR/AAV weighed 17-tons and was designated the GM M551 *Sheridan* light tank. It was not heli-transportable. It was used in Armored cavalry units in Vietnam in the 2-D mode along with M48 tanks and M113A1 AFVs[23].

During the 1950s, the U.S. Army and industry watched British and French helicopter operations in Malaya and Algeria and steadily improved their own helicopter designs and capabilities. General Gavin, as Army Chief of Operations, ordered numerous staff studies on Cavalry organizations using helicopter mobility. In tests, helicopters armed with rockets and machine guns (like the French helicopters had done in Algeria) suppressed Landing Zones (LZs) before troop carriers landed. Since there was no official Army doctrine for helicopter use, the various service branches developed their own applications of helicopters. The Army had upgraded from the Sikorsky H-19 to the Piasecki H-21 "*Shawnee*" (unofficially as the "*Flying Banana*") and the H-34 "*Choctaw*". But these were piston-engined and maintenance intensive. What was needed was a simpler, lighter engine. The Bell XH-40 Helicopter Utility, HU-1 with one of the first turbine engines was the answer. The Boeing Vertol HC-1B *Chinook* (design from Piasecki) with its giant turbine engines replaced the piston-engined Sikorsky H-37 *Mojave* as the Army's heavy lifter. Both would become legendary aircraft re-designated as the utility helicopter-1 or "UH-1" and as the Cargo Helicopter model 47 or "CH-47". By 1960, the Army had 5,000 rotary and fixed-wing aircraft! The Army was now the world's leaders in practical helicopter use but without a doctrine.

The Army Aircraft Requirements Board was created In January 1960 to establish a unified Army course and it was led by Lieutenant General Gordon B. Rogers. The "*Rogers Board*" made some improvements in procurement and planning but did not actually test hardware and aircraft to create an Airmobile doctrine for the Army. In 1961, Army Aviation units were deployed to Vietnam, with three H-21 transport helicopters to lift South Vietnamese troops into battle and a company of *Otter* fixed-wing spotter/radio-direction finding aircraft. The first turbine-engined UH-1 "*Huey*" helicopters arrived in 1962. As the Army fought with the Air Force to keep its own modest fleet of fixed-wing OV-1 *Mohawks* for Close Air Support, and *Caribou* Short Take-Off and Landing (STOL) aircraft for forward base resupply, the development of helicopter doctrine languished.

The rest of the Army had returned to 2-Dimensional warfare. However, faced with the prospects of a military conflict in the jungles of Southeast Asia, the Secretary of Defense, Robert McNamara sent a memorandum to the Secretary of the Army stating that the Army had not fully explored technological opportunities to break its ties to surface mobility. The result was the *Howze* Board formed to look into 3-D helicopter Air Assault mobility.

The Secretary urged the Army to reexamine its aviation requirements with a "bold, new look", at land warfare mobility. The reexamination, his memo said, could come only through testing; more studies on the same subject would not suffice. Moreover, McNamara directed that the testing and reporting be divorced from traditional viewpoints and policies and free from veto or dilution by conservative staff review. And then, in a stunning departure from usual OSD protocol, he suggested a number of officers and key civilians, headed by Lieutenant General Hamilton H. Howze, to manage the Army's efforts.
—-*Chopper*, Lieutenant Colonel J. D. Coleman, (New York, St. Martin's paperbacks, 1988) page 6.

The rugged Vietnam terrain and the general desire of the helicopter replace ground vehicles with helicopters led the Board to recommend helicopter acquisition/development for dismounted infantry but not ground vehicles. Consequently, the U.S. Army formed the 11th Airborne Division as the first Air Assault test-bed unit. The "*Angels*" were then disbanded and reorganized with its equipment and some of its personnel into the 1st Cavalry Division, re-designated as the world's first helicopter Air Assault unit. General Galvin observed; "*The first Division –size Air-Mobile unit (the American 1st Cavalry Division) proved in all it's major Vietnam battles that three-dimensional tactics created advantages of mobility and firepower completely offsetting the enemy's knowledge of the terrain and his ability to mass and disperse rapidly".*

Under the energetic leadership of General Harry W. O. Kinnard, the 1st Cavalry Division had created partial combined-arms team of helicopters and light troops, but with no small helicopter-deliverable AFVs like the AR/AAV. The early Air Assault concept was envisioned that the helicopter would replace ground vehicles almost completely; but as combat showed; aircraft cannot stay support the infantry indefinitely but must go back to base to refuel and rearm. Once Air Assault infantry units were landed onto the ground, they had no armored vehicles to transport and sustain them by carrying supplies or providing protection and firepower. All they had was what could be man-packed.

Before the M551 *Sheridan* was developed, officials realized that the Army needed a light, small AFV for armored reconnaissance and heli-borne fire support. Three small tracked AFVs were fielded by U.S. Army and marine units in the late 1950s (years before the *Sheridan* was fielded in 1966). These were the M56 *Scorpion* with open-topped 90mm self-propelled gun, the M114 scout vehicle and the M50 *Ontos* 6 x 106mm recoilless rifles/fire support vehicle. They were transportable by CH-47 and CH-53 helicopters, but were not well received in Vietnam because of poor automotive performance, reliability, and dangerous gasoline propulsion and were withdrawn from active service with no replacements[24].

1960s: Early *Air*-Mech-Strike with ACRs and ACAVs Highly Successful in Vietnam but then Forgotten

LTC Kris P. Thompson notes in *Trends in Mounted Warfare, Part III: Korea, Vietnam and Desert Storm*, U.S. Army Armor magazine, July/August 1998[25] that *"U.S. forces involved in the operation included 1st Cavalry Division (Air Assault), 25th Division, and the 11th ACR.* **Brilliant use of aviation and armor in mobile warfare** *led to success at the tactical level. Surprised enemy units were encircled and annihilated. Huge stocks of individual weapons, crew served weapons, ammunition, and rice were captured. The penetrating forces over-ran an extensive logistical base with a fully equipped motor park complete with grease racks and spare parts. By the end of the operation almost 10,000 tons of material and food had been destroyed and over 11,000 enemy Soldiers killed".*

The idea of a fast-moving combined-arms organization with light tracked AFVs did find a home in the separate Armored Cavalry Regiments (ACRs). In 1957, the 11th Armored Cavalry *"Blackhorse"* Regiment (ACR)[26] was assigned to Germany as part of the NATO Forces protecting the border from communist aggression until it returned to the United States in 1964. In March of 1966, the unit was alerted for movement to the Republic of Vietnam. It then began redesigning its equipment for a new type of warfare. Learning from the Vietnamese experiences at the 1963 battle of *Ap Bac*, additional armor and two more (7.62mm) 30 cal. Medium Machine Guns were added to their M113A1 Armored Personnel Carriers (APCs—which already had one 50 cal. Browning Heavy Machine Gun used by the Track Commander) Protective gun shields for the crew and track commander were also added.[27] Other ACAVs had 106mm Recoilless Rifles and 7.62mm mini-Gatling guns mounted. The result was a very rapid all-terrain fighting vehicle, which could deliver devastating firepower. The Armored Cavalry Assault Vehicle or "ACAV" had an excellent reputation and was rated the most terrain mobile vehicle of the war. The *Blackhorse* troops arrived in South Vietnam on September 7, 1966, and soon engaged the enemy with tanks, ACAVs, artillery and helicopters. The success of the ACAV in battle prompted the U.S. Army and its allies—particularly the Australian Armored Regiment—to convert other M113s in other units in a similar fashion. Brian Ross citing General Don. A. Starry in *Armoured Combat in Vietnam*, and Simon Dunstan's, *Vietnam Tracks* concludes:

> *...mounted combat came to the fore for infantry in the form of the ACAV (Armoured Cavalry Assault Vehicle). Until Vietnam, the U.S. Army's doctrine had been that infantry units should dismount before assaulting an enemy position. However, as the ARVN discovered, this meant that when facing the massive amounts of firepower that the NLF or VPA could bring to bear during a firefight, the infantry was exposed to needless casualties, as well as losing the momentum of the attack. Indeed it was the ARVN which pioneered the use of mounted tactics from APC's when they first deployed the M113 in 1962. They were also the first to discover the need for increased firepower on the vehicle by mounting an extra .30 Cal. MMG beside the commander, fired by an exposed prone Soldier lying on the roof of the vehicle. Perhaps more importantly, they also discovered the vulnerability of the exposed track commander when manning the pintle*

mounted .50 Cal. HMG during the battle of Ap Bac where 14 out of 17 commanders became casualties.

The U.S. Cavalry units...took to the idea and improved upon it by creating the ACAV. They added armour around the commander and a gun shield for the .50 Cal., provided two extra M60 GPMG's each athwart the roof hatch (protected by shields) and installed an M79 Grenadier inside the troop compartment, firing through the roof hatch to provide close support. The result was a vehicle, which was able to go where tanks weren't, by virtue of its lighter weight and ground pressure, packed considerable firepower and was agile and reasonably well armoured. The result, when coupled with the aggressive leadership and tactics of the U.S. Cavalry's commanders was highly effective by all accounts.

The 11th Cav's main operational area was the province around Saigon and up to the Cambodian border. The unit clearly demonstrated it's rapid mobility when Saigon came under siege during the 1968 Tet Offensive. James Arnold writes:

> Half an hour after the opening barrage, the 2/47th Battalion (Mechanized) began a speed march from Bear Cat toward Long Binh. At first light, the 2d Battalion, 506th Infantry airlifted into Bien Hoa air base. The 11th Armored Cavalry Regiment, the Blackhorse regiment, made a 12-hour forced march to arrive at Long Binh during the day. Once in position, the multiple machine guns of the mechanized unit's APCs shot apart all Viet Cong attacks.

Note that the key to this success was the M113 ACAV's ability to get OFF THE ROADS and go cross-country to hit the enemy where he was not ready.

> Nearing the air base he spotted hundreds of enemy soldiers belonging to the 274th VC Regiment lining Highway 1, apparently deployed to stop any relieving column. Exploiting its mobility, the cavalry left the highway and drove a parallel route. The ACAV's

machine guns shot up the unsuspecting enemy from the rear and finally reached the air base...its presence at Bien Hoa provided the narrow margin between victory and defeat. Along with the 2d Battalion, 506th Infantry, it repulsed all assaults.

A Cavalry officer wrote:

> Saigon, Bien Hoa and Long Binh were literally ringed in steel...Five Cavalry squadrons had moved through the previous day and night, converging on the Saigon area. When dawn broke, they formed an almost continuous chain of more than 500 fighting vehicles...we actually cheered...from that morning the outcome was never in doubt. We knew that our enemy could never match our mobility, flexibility and firepower
>
> —*Tet Offensive 1968, Turning Point in Vietnam*, James R. Arnold[28]

In July of 1968, Colonel George S. Patton III, the son of the WWII hero, assumed command of the 11th ACR and soon applied his expertise in armored combat and **moved the armor off the roads and into the jungles in search of the enemy**. So successful was the unit's search and destroy missions within the enemy's main supply routes between Cambodia and Saigon, that the enemy could no longer move freely and was forced to seek sanctuary inside neutral Cambodia.

Essentially in Vietnam, the U.S. Army used light tracked M113 armored vehicles and helicopter gunships, troop carriers and scouts to find, fix and gain a temporary mobility superior to that of the enemy, but couldn't combine the two forms into an integrated air-delivered *Air*-Mech whole due to a lack of a helicopter-transportable small AFV. General William C. Westmoreland, Commanding General U.S. Forces, Vietnam, states in *Tet Offensive 1968, Turning Point in Vietnam*, James R. Arnold, p. 36:

"The ability of mechanized cavalry to operate effectively in the Vietnamese countryside convinced me that I was mistaken in a belief that modern armor had only a limited role in Vietnam"

The M551 *Sheridan* withdrawn from Army service in the 1970s except in the 82d Airborne Division. The U.S. Army was without a turret-equipped, high power-to-weight ratio light tracked recon vehicle to scout ahead for the 2-D main body and certainly without a small AFV that could accompany 3-D helicopter Air Assaults for fire support and/or troop transport. The M151 1/4 ton 4-wheel drive jeep was an excellent mount that could fit inside helicopters, be easily airdropped and could mount machine guns, 106mm Recoilless Rifles and TOW anti-tank missiles. The drawback was its unarmored configuration and vulnerability to enemy fire, and flammable gasoline gas tank. The smaller M274 *Mule* 4x4 had the same attributes and vices but was mechanically unreliable. Both vehicles were used widely throughout Vietnam; but they could not break brush and move cross-country like the light tracked M113 could.

The British Army in the '70s, fielded an 8-ton small *Scimitar/Scorpion/Spartan* tracked AFV family that could be used in heliborne operations they did fly them into action in the 1982 Falklands islands war despite a lack of transport helicopters. Most of the invasion force's RAF CH-47s were lost when their ship, the *Atlantic Conveyer* was sunk by an *Exocet* anti-ship missile. This lack of Airmobility forced the British Paratroopers to walk across the islands. *Scorpion/Scimitar* light tanks with their low ground pressure were able to drive over the soft boggy terrain of the Falklands and were employed in attacks to provide critically needed fire support to Paratrooper and infantry units resulting in victories with light casualties[29]. In 1999, British Paratroops helo-lifted *Scimitar* light tanks with CH-47s into Kosovo to bypass unexploded ordnance and mined roads.

In the 1980s, the U.S. Army fielded 33-ton infantry fighting vehicles (*Bradley's*) and 70-ton turbine-engined *Abrams* tanks for its combat units. M113 light tracked vehicles remained with engineer, C2 and combat support units. These were up-graded to M113A3 standards to keep pace with the M1/M2 in open terrain. Reconnaissance units were equipped with either the M3 Cavalry Fighting Vehicle or the HMMWV. Infantry units were either "heavy" (equipped with *Bradley's)* with reduced dismounting infantry strength, or "light" (on foot delivered by parachute, helicopters or trucks). Since Vietnam, Army light forces for the most part due to low foot mobility have had to land on or very close to objectives and "hang on" until heavy forces arrive unless the enemy threat was minimal.

In 1971-74, the Army's TRICAP (triple capability) 1st Cavalry Division using an Airmobile Infantry Brigade as a force with tactical and operational mobility, an Air Cavalry Brigade with ATGM firing Cobra helicopters as the 3-D element and an Armored Brigade as a 2-D maneuver element showed promise but was dropped when the command and control technology of that time was not able to synchronize all the elements.

The force structure was not deemed armor heavy enough to prevail in a 1973 Arab-Israeli-type tank war against the Soviets in Europe.

3-D *Air*-Mech Maneuver by U.S. Army Airborne Stalled

The most important innovation for "high intensity" ground war in the 20th Century was the creation of the mechanized infantry Division with each infantry battalion equipped with light (11-ton) M113 Armored Fighting Vehicles (AFVs). This did not include equipping the 82d Airborne Division to create an *Airborne Mechanized* infantry capability even though M113s could have been airdropped by USAF C-130 tail-ramp equipped aircraft dating back to the 1950s. An Airborne unit with "tracks" is not unprecedented. In the 1960s and early 70s, the 1st Brigade of the 8th Infantry Division had the 1st and 2nd Battalions of the 509th Airborne Infantry (located at Mainz, Germany) which were mechanized with M113s.

During the 1956 Suez crisis, French AMX-13 (90mm-105mm gun equipped) light tanks were effectively used for rapid intervention by both the French and Israelis. The Israelis used their combined *Air*-Mech operation to take the *Mitla* pass[30]. AMX-13s could have been purchased by the U.S. Army Airborne to fill out a complete Airborne combined-arms team with the M113s carrying Paratrooper infantry as some authors advocated at the time[31]. The Russian Airborne fielded ASU-57 self-propelled gun in the late 1950s. These semi-armored tracked carriers would be used by Russian advisors in the World's first combat helicopter *Air-Mech*anized assault in the Ethiopian-Somali Ogaden War.

The U.S. Army could have had *Airborne*-Mech-Strike capabilities in the 1960s using the M113 long instead of having the Russian Airborne (VDV) take the lead in perfecting the capability with its ASU-family of assault guns and BMD family of 8-ton AFVs used by her Airborne Divisions to subdue Czechoslovakia and Afghanistan[32].

Russian Airborne 3-D *Air*-Mechanized Maneuver in Combat

After World War II, the Soviet Army re-examined its concepts for *Air*-Mechanization and began building ground mobility into its Airborne units. By 1953, self-propelled, lightly-armored tracked assault guns were part of Airborne units. Trucks and gun tractors were also deployed for further ground mobility. By 1968, the BMD [B*oevaya Mashina Desantnaya*] parachute-delivered armored personnel carrier entered Soviet service. The Soviet Airborne Division had become fully ground mobile. Over the years, the Soviet Army developed a family of armored, tracked vehicles for their Airborne forces. Infantry personnel carriers, anti-tank missile carriers, air defense guns, artillery pieces, assault guns, command vehicles, communications vehicles, recovery vehicles, and reconnaissance vehicles were all developed for air delivery.[33] The seven peace-time Soviet Airborne Divisions were fully-equipped and 100% ground mobile when they hit the ground. The Soviets saw Airborne forces as an operational force for deep battle which would leapfrog over nuclear-strike zones and maneuver behind NATO's rigid linear defense. *Air*-Mechanization would convert a war on the northern European plain to a non-linear contest in which the Soviets felt they had the advantage. They prepared for deep battle and *Air*-Mech would play a vital role in it. *Air*-Mech and the Operational Maneuver Group (OMG) were designed as complimentary components of deep battle.[34]

The first real test of the Soviet *Air*-Mech concept was in Czechoslovakia in 1968. Czechoslovakia was trying to liberalize and perhaps break away to become a neutral buffer state between NATO and the Warsaw Pact. The Soviet leadership determined that Czechoslovakia would remain a part of the communist bloc and acted against the Czechoslovakian national command and control. Soviet Airborne forces air-landed in major Czechoslovakian airfields, mounted their vehicles and sped off to seize garrisons, key bridges, communications centers, and government buildings. Czechoslovakia's government was firmly under the Paratroopers' control long before the ground columns drove into Prague. Within days, Soviet tanks were in the streets of Prague and the leaders of the movement, including President Alexander Dubcek in house arrest. The *Air*-Mech operation was so rapid that resistance was futile and the country fell with little bloodshed.

The Soviets studied American use of helicopters in Vietnam and Israeli use of helicopters during the six-day war in the Sinai.[35] They began developing their own heliborne concepts. However, the Soviets felt that helicopters represented more than transport and gun platforms. The Soviet heliborne force needed to become a lethal Air Assault *Air*-Mech maneuver force. Air Assault forces needed significant firepower and ground mobility. Consequently, the same family of armored vehicles which were assigned to the Airborne were also assigned to the Air Assault units. Helicopter gun ships and armored vehicles were integrated into a lethal force. The Soviets developed heavy-lift helicopters which could carry the vehicles into Air Assault insertions. Inserted forces learned to maneuver in conjunction with helicopter gun ships. The Soviets eventually fielded nine Air Assault brigades. Both the Soviet Airborne and Air Assault community studied and practiced *Air*-Mechanization in a variety of Eurasian exercises, but the next combat testing of their concept occurred in Africa.

The Horn of Africa is a strategic chokepoint controlling entry into the Red Sea and the southern end of the Suez canal. Djibouti and Yemen control the chokepoint while Ethiopia, Somalia and Ethiopia flank the approaches to the chokepoint. Throughout the 1950s and 1960s, Ethiopia was an ally of the United States. U.S. military advisers helped train the Ethiopian Army and maintained vital communications sites in Ethiopia. In 1963, the Soviet Union supported Ethiopia in the Somalian-Ethiopian land dispute in the Ogadan desert and further backed Ethiopia in its fight with Eritrean separatists. However, at the same time, the Soviet Union backed

Ethiopia's bitter enemy, Somalia. Soviet advisers were stationed throughout the Somalian Army and Somalia provided a naval base to the Soviet navy at Berbera. Somalia supported the Eritrean separatist movement in Ethiopia.[36]

In 1974, an Ethiopian revolution overthrew the emperor, forced out the United States advisers, and began the creation of another Marxist-Leninist state. Soviet advisers, wearing Ethiopian uniforms without any rank or insignia, arrived in Ethiopia in February 1977. Somalia invaded the Ethiopian Ogaden desert in July of that same year and the Soviets found themselves advising both sides of the conflict. In November 1977, Somalia expelled all her Soviet advisers, who transferred to Ethiopia. By the end of 1978, there were 3,000 Soviet advisers, and 18,500 Cuban combat troops in Ethiopia. The Soviets shipped T-54 and T-55 tanks, 130mm artillery pieces, air defense systems, MIG-21 and MIG-23 fighter aircraft, helicopters and military vehicles to her new ally. The Cubans were even better armed with new T-62 tanks and infantry fighting vehicles.[37] Soviet General of the Army, V. Petrov and his advisory staff helped the Ethiopians plan and mass forces against the Somalian force in the Ogaden. The Cuban forces spearheaded the breakthrough while Ethiopian forces, and an artillery Brigade from South Yemen, advanced on the flanks.[38] Following a 20-minute artillery preparation, the Cuban main attack penetrated the Somalian defense. The Somalis fought a withdrawal and after over a month's hard fighting, were holding strong positions in a mountain chain. The Soviet/Cuban/Yemeni/Ethiopian force had air superiority, but was stymied as long as the Somalis held the mountain passes. At this point, a regiment of Soviet Mi-6 heavy transport helicopters lifted an armored force of 70 Cuban BMD-1 and/or Air Assault vehicles to the rear of the mountain positions. Then the Cubans launched a tank-heavy flanking attack around the mountains against the town of Jijiga while the *Air*-Mech force attacked Jijiga from the other direction. The sudden appearance of a Cuban *Air*-Mech force in the rear area, combined with the Cuban flanking attack, unhinged the Somali defense.[39] The communists advanced some 350 kilometers and pushed Somalian forces out of Ethiopia.[40]

Vietnam Armor combat veteran Ralph Zumbro writes in his book, *The Iron Cavalry* that the Russians used ASU-57s in the

How *Air*-Mech-Strike Builds the Bridge to the Future

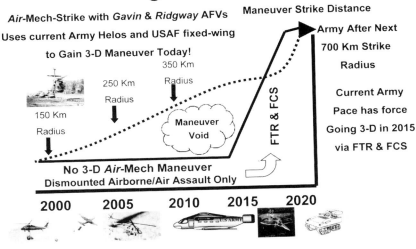

How *Air*-Mech-Strike builds bridge to the future
(Major Charles Jarnot and Carol Murphy)

Russian ASU-57 Assault guns and Paratroopers on the march
(U.S. Army)

epic *Air*-Mech assault in 1978[41], so we are unclear if the assault was all BMD, all ASU-57 or a combination of both.

The next opportunity that the Soviets had to test the *Air*-Mech concept in combat was in the Soviet-Afghan War of 1979-1989. Building on their practical experiences, the Russians were ready when the call came to subdue the capitol city of Kabul, Afghanistan in 1979.

The Soviets deployed the 40th Army which contained several *Air*-Mech units: the 103rd Airborne Division, the 345th Separate Parachute Regiment, the 56th Air Assault Brigade, the 15th Spetsnaz Brigade, the 22nd Spetsnaz Brigade, the 66th Separate Motorized Rifle Brigade and the 70th Separate Motorized Rifle Brigade. Three helicopter Regiments provided lift to these forces. The initial use of *Air*-Mech forces was brilliantly executed. Leading the invasion, the 103rd Airborne Division and 345th Separate Parachute Regiment air-landed at Kabul and Bagram airfields respectively. The Paratroopers drove off the aircraft in their armored personnel carriers to seize the airfields, key government buildings and key government officials. Within hours, the *Air*-Mech forces controlled the government of Afghanistan. The 56th Air Assault Brigade seized Kunduz airfield and city in a similar fashion.[42]

Control of the Afghan government did not mean control of the Afghan countryside. The Soviets found themselves trying to control an area five times larger than South Vietnam with one-fourth the number of troops committed by the United States against the Vietnamese communists. Motorized rifle forces ended up securing the cities, garrisons and limited road networks while the *Air*-Mech forces took the fight to the Afghan *Mujahideen* [Holy warriors]. The terrain of Afghanistan is forbidding. Over 49% of the total land area is over 6,000 feet in elevation. The rugged Hindu Kush mountains, which are an extension of the Himalaya mountain range, cut the country in half and reaches over 21,000 feet in height. The unforgiving Registan desert covers the southern third of the country. The *Air*-Mech forces had to contend with the guerrillas in the inhospitable part of the country. Airborne forces quickly discovered that helicopter landings were preferable to parachute landings in the mountainous terrain and conducted their insertions by helicopter. There were a limited number of helicopters available and heavy-lift helicopters were particularly in demand. Helicopters cannot lift heavy loads above 13,000 feet heights at which the guerrillas were often found. The fine sand common throughout Afghanistan found its way into the helicopter mechanisms. This greatly increased wear and tear and increased maintenance time. The landing zones in the mountains often could not accommodate the *Air*-Mech vehicles. The *Air*-Mech forces often found that they had to leave their vehicles behind. However, they usually had the vehicles move *en masse* to a point where they could meet the dismounted troops or support them by fire.[43]

After the Soviet withdrawal from Afghanistan, the Soviet Airborne was used to suppress demonstrators and try to shore up a collapsing Soviet Union. Russian Paratroopers have seen action in Georgia, Ossetia, Dagestan and Chechnya, but only occasionally in an *Air*-Mech role. In 1999, immediately after the Kosovo cease-fire was agreed upon by all the participants, the Russian Airborne nearby in Bosnia seized the initiative and took control of the capitol province's international airport in a light mechanized vehicle *coup de main*.

The American Airborne Achieves a 3-D *Air*-Mech Maneuver First

While the Russians were fielding entire Airborne Divisions centered around the parachute airdrop of light AFVs, the U.S. Army became the first to actually do it in combat. The 82d Airborne's 3d Battalion/73d Armor turned the outcast 17-ton *Sheridan* M551 light tank into a success and a legend. They upgraded and modernized their mounts and by paradropping 10 of these tanks from C-141B *Starlifter* jet aircraft into night combat during the invasion of Panama in 1989.[44] Other *Sheridans* were air-landed into Panama days earlier by C-5Bs. They landed in the middle of the night, to maintain secrecy, exploiting the light AFV's air-transportability for maximum tactical gain. Special Forces units used M113 APCs to help storm Modelo prison to rescue American Kurt Muse. Mechanized Infantry in M113s supported attacks to subdue the center of enemy resistance, the buildings known as "*the Comandancia*". The *Sheridan*'s

152mm main gun shock action was devastating to the thick concrete walls of the enemy's headquarters buildings *La Comandancia*.[45] Thanks to bold 3-D maneuver, U.S. forces were able to overwhelm quickly and subdue the Panamanian Defense Forces. U.S. forces had low casualties and captured strongman Manuel Noriega after he was surrounded by *Sheridans* at a Catholic church. Months later, during Operation DESERT SHIELD, the 3d Battalion/73d Armor AIR deployed with their M551 *Sheridans* to Saudi Arabia as the initial U.S. Army armor force on the ground to deter Iraqi aggression.

In 1997, the 3/73d Armor Battalion of the 82d Airborne Division was disbanded and its *Sheridan* light tanks retired and the M8 Armored Gun System (AGS) canceled[46]. The U.S. Army is now without an armored AMS capability. The clandestine use of the *Air*-Mech-Strike concept by flying in Army Special Forces in vehicles delivered by CH-47D helicopters to locate and destroy *Scud* missiles during the 1990-91 Gulf War proves its viability in a Major Theater of War (MTW) setting and continues to the present day with the acquisition of newer, though thinly armored vehicles for U.S. Army Special Operations Forces (SOF). The All-Terrain Vehicle (ATV) is also widely used by Army SOF because of its small size and air-transportability for similar mission tactics.

At the present time, the conventional U.S. military does not have an integrated *Air*-Mech-Strike 3-D maneuver capability. The only conventional forces America can "**force-entry**" by air are on foot with at best thin-skinned wheeled vehicles. Other Armies with a fraction of our budget already have air-transportable tracked armored fighting vehicle *Air*-Mech-Strike capabilities: Russia, Great Britain, Germany, Israel, Italy, South Africa and India[47]. The recent announcement by the Army CSA in October 1999 creating Medium-sized Brigades marks the end of the Cold War confrontation "*Land Battleship*" Army and the beginning of the "*power projection*" U.S. Army that *Air*-Mech deploys by aircraft. This is a needed change for the U.S. Army.

2000 and the Future: Movement is the First Battle of any War

B.H. Liddell-Hart *said; "Throughout the ages, effective results in war have rarely been attained unless the approach has had such indirectness as to ensure the opponent's unreadiness to meet it. The indirectness has usually been physical and always psychological.*" Sensors and human intelligence increasingly cover the modern battlefield. Whatever the enemy observes can be instantly reported and exploited. Even Somali gunmen had "*walkie-talkies*" and command-detonated land mines.[48] In the past the U.S. could add armor to fighting vehicles to ignore many threat weapons. However, signatureless, top-attack anti-tank missiles have become more available to our enemies and negates our invulnerability to massed missile attacks and may make maneuver impossible through belts of enemy obstacles, mines and ATGM fields of fire all inter-connected by information technology means.[49] The U.S. Army must adapt *Air-Mech-Strike* to regain the "*edge*" in maneuver warfare by having a 3-Dimensional maneuver capability to unhinge defenses via avoiding them and gaining positional advantage.

Structurally, today's U.S. Army is not much different than the force we had at the end of the Cold War—a mix of light units with no AFVs and heavy units with very heavy AFVs. One force experienced in 3-Dimensions to get positional advantage, the other comfortable in 2-Dimensions.[50] The U.S. Army is being changed into "*Army XXI*" with a focus on major upgrades in Command, Control, Communications, and Computers (C4) and improved Intelligence, Surveillance, and Reconnaissance (ISR) systems but not unified to create a multi-dimensional maneuver capable force.[51] Below is a list of current active-duty U.S. Army combat units and their classification by "*weight*".

Light 3-D Forces: They Can Fly into Positions of Advantage; Disadvantage: No AFVs for Highly Mobile 2-D Maneuver upon Arrival

82d Airborne Division "*All Americans*"
10th Mountain Division "*Mountaineers*"
25th Light Infantry Division "*Tropic Lightning*"
2d ACR "*Toujours le Pret*"
172d Infantry BDE "*Arctic Warriors*"

1-501st Parachute Infantry Regiment (PIR) "*Geronimos*"
1-508th PIR "*Red Devils*"
1-509th PIR OPFOR JRTC "*Gingerbread Men*"
75th Ranger Regiment "*Rangers*"
Numbered Special Forces Groups "*Green Berets*"

Medium Forces: Helicopters: No Heli-transportable AFVs Configured to Actually Fight

101st Air Assault Division "*Screaming Eagles*"
Owns the world's largest fleet of helicopters assembled in one location.
11th ACR "*Blackhorse Regiment*"
Has lightweight M113A3 OSV and M551 AFVs but as National Training Center OPposing FORces (OPFOR), their mounts have been visually modified to look like Russian made equipment (VIS-MOD) and are more training aids than fighting means.

Heavy 2-D Forces: Many Heavy AFVs, Disadvantage: no 3-D Maneuver, can't Maneuver Freely in Unimproved Third World Areas

1st Infantry (Mechanized) "*Big Red One*"
1st Armored Division "*Old Ironsides*"
1st Cavalry Division "*The First Team*"
2nd Infantry Division "*Second to none*"
3rd Infantry (Mechanized) "*Rock of the Marne*"
4th Infantry (Mechanized) "*Ivy Men*"
National Guard 7th Infantry Division (Mechanized) "*Bayonets*"
National Guard 24th Infantry Division (Mechanized) "*Victory Division*"
3d Armored Cavalry Regiment "*Brave rifles*"

Combined 2-D/3-D Forces: Best of Both Attributes?

The 10-Division U.S. Army of 2000 onwards, has seven "heavy" armored or mechanized infantry Divisions on one end of the spectrum and three air-transportable light Divisions on the other end. The 2d Infantry Division, 101st Air Assault Division and

11th ACR at Fort Irwin, California lie in the middle. While the 101st Airborne (Air Assault) Division doesn't have AFVs, it is logistically equivalent to a heavy Division and difficult to move long distances by USAF aircraft due to the volume of their large numbers of helicopters and mostly unarmored wheeled vehicles.

The 2d Infantry Division *"Second to none"*, on the other hand, is very interesting and a great model for the future Division. Situated in a volatile environment, they have a mix of Air Assault light infantry Battalions, heavy mechanized infantry and tank battalions in order to enhance their battlefield *"flexibility"*.

The Current Army Transformation Initiative

The result of the delay of the current Army Cold-War confrontation-based heavy AFV forces to deploy in a timely or effective manner to conflicts like Bosnia (1996) and Kosovo (1999) has caused the Chief of Staff of the Army (CSA), to announce the beginning of the less than 20-ton *"Medium"* Interim Brigade Combat Team force structure designed to take the Army from the near present until when the Future Combat System (FCS) is fielded.[52]

Meanwhile, while the U.S. Army began considering restructuring possibilities, thousands of miles away, a daring operation led by an American ally, showed the world how air-delivered forces with light tracked armored fighting vehicles can be the decisive force in modern warfare, getting there *"fustest with the mostest"*.

East Timor: Australian Light Tracked M113s Flown in by C-130s

A vivid example of the efficiency of the light tracked M113, as an air-delivered contingency fire support platform was the East Timor crisis at the end of 1999. As C-130s began airdropping food to refugees hiding in the mountains, they began landing at the international airport. Immediately after landing, camouflaged, machine gun turret-mounted M113 tracked AFVs were rolling off the ramp of cargo aircraft. Within hours, these AFVs were leading Australian infantry through the streets to put an end to the violence. The "thugs" in army uniforms and "paramilitary" in civilian clothes knew with the sound of tracks coming their way that the peacekeeping force meant business.

Lieutenant Colonel Wolfgang Mettler, a former German Army *Wiesel* Airborne Battalion Commander concludes in the January-February 1995 issue of U.S. Army Infantry magazine;

> *"Armament and equipment fundamentally determine the principles of employment. On the eve of World War I, the disregard for the machinegun that was coming into use turned maneuver warfare into trench warfare, but at the end of the war the emergence of the tank made maneuver again possible in spite of machinegun fire. Following World War I, visionary military thinkers formed the armor into an element that made possible extensive, enveloping movements by means of the massing of forces, movement, and firepower. the early days of World War II saw the classical formation and employment of armor in combination with the air forces and large follow-on motorized units. Today, we have in essence an armor force that is highly evolved technologically, tactically and operationally, but whose capability for rapid mobility and momentum can be clearly restricted by modern countermobility measures...In the context of such developments, a unit that can rapidly bypass these and other impediments—while retaining its combat capability—through the use of the dimension of Air-Mobility achieves a new significance."*

---

[1] CSA Gen Shinseki's speech to AUSA (Fort Bliss, Texas, U.S. Army Air Defense Artillery Magazine online; October 1999) *http://147.71.210.2/adamag/Oct99/CofS.htm*

[2] Ralph Zumbro, *Iron chariots* (New York, Simon & Schuster Pocket Books, 1998), pp. 147-174

[3] Ian Hogg, *Great Battles of World War 2* (New York, Double Day, 1987), pp. 9-24

[4] For pictures of the German Airborne's WWII *Air*-Mech-Strike combination: *http://www.geocities.com/Pentagon/Quarters/2116/hptll.htm*

⁵ Brigadier General Maurice A. J. Tugwell, British Army, *Is the day of the Paratrooper over?* (Fort Leavenworth, Kansas, U.S. Army Military Review magazine, March 1977), pp. 75-83 http://www.geocities.com/Pentagon/Quarters/2116/dayofparatrooper.htm

⁶ Richard Simpkin, *Race to the Swift: Thoughts on Twenty-First Century Warfare*, (London: Brassey's Defence Publishers, 1985) pp. 104

General Doctor F. M. von Senger und Etterlin, *Die roten Panzer: Geschichte der sowjetischen Panzertruppen von 1920-1960*, (Munich: Lehmanns Verlag, 1963)

⁷ Mikhail N. Tukhachevskiy, <u>Novye voprosy voyny</u> [*New Issues of War*], material written by the author in the 1931-1932 period and unpublished until 1962. Material is collected in G. I. Os'kin and P. P. Chernushkov (ed), <u>M.. N. Tukachevskiy isbrannye proizvedeniya</u> [*Selected Works of M. N. Tukachevskiy*], (Moscow: Voyenizdat, 1964) Volume 2, p.184.

⁸ D. S. Sukhorukov, P. F. Pavlenko, S. M. Smirnov, V. F. Margelov, I. I. Lisov, Ya. P. Samoylenko, and V. I. Ivonin, <u>Sovetskie vozdusho-desantye</u> [*Soviet Airborne*], (Moscow: Voyenizdat 2nd edition, 1986), p.12.

⁹ Ibid, 14

¹⁰ Interview by LTC Lester Grau with Jacob W. Kipp, who, as always, has provided valuable input to this piece

¹¹ David M. Glantz, <u>A History of Soviet Airborne Forces</u>, (London: Frank Cass Publishers, 1994) pp. 7-8.

¹² Ibid., 7-9.

¹³ Ibid., 11-12.

¹⁴ Ibid, 14-17

¹⁵ Bill Gunston, *Aircraft of the Soviet Union*, (London: Osprey, 1983), pp. 40-41.

¹⁶ Glantz, 68.

¹⁷ Sukhorukov, 387-391.

¹⁸ Col (Major) Michael Kazmierski, *United States Army Power Projection in the 21st Century: the conventional Airborne forces must be modernized to meet the Army Chief of Staff's strategic force requirements and the Nation's future threats* (Fort Leavenworth, Kansas, U.S. Army Command and General Staff

College Master's Thesis, Defense Technical Information Center File # AD A226 216), pp. 13-53

http://www.geocities.com/Pentagon/Quarters/2116/airbornetoc.htm

[19] George S. Patton, *War as I knew it* (New York, Houghton Mifflin Company, 1995), Chapter IV,

Stuck in The Mud, pp. 163

[20] William B. Bruer, *Drop Zone Sicily: Allied Airborne Strike, July 1943* (Novato, California, Presidio Press paperback, 1997)

[21] LTC J.D. Coleman (retired), *Choppers* (New York, St. Martin's paperbacks, 1988), pp. 2-10

...the precursory decision for air-mobility occurred in 1952, when the Army proposed forming twelve helicopter battalions. This decision was made even though a practical troop-carrying helicopter was still unproven and helicopter tactics and techniques existed only in the minds of a few visionaries...While Chief of operations for the Army in 1954, General James M. Gavin, the legendary World War II Paratroop commander, ordered a series of staff studies to design hypothetical cavalry organizations around the potential of the helicopter.

Korean war web site: http://rt66.com/~korteng/index.htm

[22] Lieutenant General Harold G. Moore; telephone interview with Michael Sparks; January 15, 2000

[23] Dr. Robert Cameron, U.S. Army Armor Center Chief Historian, *American Tank Development during the old war: maintaining the edge or just getting by?* (Fort Knox, Kentucky, U.S. Army Armor magazine, July-August 1998) pp. 34-36

[24] Ralph Zumbro, *Tank Aces* (New York, Simon & Schuster Pocket Books, 1998), pp. 260-261

"*Our scout M114 vehicles had proven disastrously unreliable and had to be replaced with another model, the M113, which was proving itself in the Mekong Delta in the capable hands of the ARVN cavalry*".4

[25] LTC Kris P. Thompson, *Trends in Mounted Warfare, Part III: Korea, Vietnam and Desert Storm* (Fort Knox, Kentucky, U.S. Army Armor magazine, July/August 1998)

[26] Official 11th ACR web site: http://www.11thcavnam.com/history.htm

[27] Ralph Zumbro, *Tank Aces*, pp. 260-261

Neil Sheehan, *Bright Shining Lie* (New York, Vintage Books, 1989) Refer to center pictures section for map drawn of the battle of *Ap Bac* and pictures of ARVN Paratroopers descending in wrong position to win the battle after the M113 attack was beaten back due to exposure of Track Commanders to enemy fire

[28] Brian Ross, *The Use of Armour in the Vietnam War*: http://www.lbjlib.utexas.edu/shwv/articles/arm-faq.htm

General Don A. Starry in *Armoured Combat in Vietnam*, (Poole, England: Blandford Press, 1981) p. 27

Simon Dunstan's, *Vietnam Tracks,* (London: Arms and Armour Press, London, 1982) p. 39

James R. Arnold, *Tet Offensive 1968, Turning Point in Vietnam* (London, Reed Consumer Books for Osprey Publishing LTD, 1990) , pp. 50-54

[29] BG Julian Thompson; *No Picnic: 3 Commando Brigade in the South Atlantic: 1982*

http://www.amazon.com/exec/obidos/ASIN/0006370136/qid=953744929/sr=1-6/103-6593991-3811860

[30] Sir Robert Thompson, *War in Peace: conventional and guerrilla warfare since 1945* (New York, Harmony Books, 1981), pp. 90-107

[31] Richard M. Ogorkiewicz, *Airborne Armor* (Fort Leavenworth, Kansas, U.S. Army <u>Military Review,</u> August 1959)

In contrast, the major problem since World War II has been the **lack of suitable tanks**. The standard United States C- 82 tactical transport of the post-World War II period was quite capable of carrying the 15,600 pound M22, but the latter, and the equally undergunned British *Tetrarch*, were already virtually obsolete when they" were first used in Airborne operations. The M24 light tank was more powerful and at 38,750 pounds could be carried in the C-124 *Globemaster* strategic transport, but in the late forties its 75mm gun armament no longer was adequate while its more powerfully armed successor, the M41, ruled itself out from the Airborne field by weighing 50,000 pounds.

Only the French Army had the foresight in 1946 to conceive a well-armed air-transportable tank— the AMX-13. Although too heavy for such tactical transports as the C-119 and C-123, and the French Nord 2501 *Noratlas*, the AMX has been carried successfully in the Bregurt 765 *Saltara* and is within the carrying capacity of such more recent transports as the C-130 *Hercules* and the British *Lleucrlcy*. Thus by the mid-fifties the AMX acquired the distinction of being the only battle-worthy air-transportable tank in service. It was only after the prototypes of the AMX appeared that the somewhat similar T92 air-transportable tank was designed in the United States. A little earlier came the 1949- specified T101 Airborne self-propelled gun, since standardized as the M56, At about the same time the Soviet Army also introduced into service an air-transportable 57mm self-propelled anti-tank gun.

The adoption of the M56 and its integration into the Airborne divisions is a welcome step forward in the direction of air-transportable mechanized weapons. But for all its virtues, the M56 barely touches upon the numerous possibilities in this field and its development was none too early.

The slow progress since World War II in the field, of air-transportable mechanized weapons has been due largely to an "unwillingness to depart" from the established trends toward "bigger and better" tanks and to reconsider the entire problem of armored vehicles in terms of the basic need of Airborne forces for mobile weapon power. A few Airborne enthusiasts advocated the development of lightweight air-transportable tanks, but in general little were done about bringing armored vehicles into Airborne operations. It was converted to such far-fetched schemes as the transport of 100,000-pound medium tanks of the M46 class in experimental giants of the C-99 type, instead of trying to develop lightweight vehicles within the 16,000-pound carrying capacity of the standard assault transports.

Admittedly, the problem of developing such lightweight air-transportable commonly held that lightweight lightly armored combat vehicles would be at too much of a disadvantage in relation to the conventional types which could afford much thicker armor. In consequence, the two organic tank battalions of the United States Airborne divisions were equipped with relatively heavy standard- type tanks which made them surface-bound and perforce relegated them to the ground follow-up echelon. As a further consequence of this state of affairs, in the late forties, attention was directed towards air-delivered vehicles; it was not easy but the difficulties were not insurmountable.

A solution to the problem of mounting sufficiently powerful conventional guns in light vehicles appeared as early as the closing stages of World War II in the shape of recoilless guns. Some of the potentialities of recoilless gun armed armored vehicles have been indicated since by the *Ontos* which has been adopted by the united states marine corps and which has been successfully parachuted from C-130 *Hercules*. More recently, even greater opportunities for the development of powerfully armored lightweight vehicles have appeared with the introduction of short-range surface-to- surface guided missiles, such as the French SS-10 and SS-11. With armament of this type capable of penetrating 16 to 19 inches of armor, light air-transportable vehicles are feasible which can destroy the most heavily armored hostile tanks and which thus can provide the answer to the most dangerous threat that Airborne units have had to face so far. At the same time, the armor-piercing performance of the Anti-Tank Guided Missiles (ATGMs) re-emphasizes the futility of trying to develop heavily armored air-transportable tanks while the case for light armored vehicles is further reinforced by the needs of the missiles themselves. For optimum effect the short-range guided missiles need to be integrated into mobile weapon systems and this calls for highly mobile launching platforms such as suitably designed combat vehicles.

Lightly armored fully tracked vehicles could weigh as little as 4,000 pounds. Yet in spite of their lightweight, such vehicles could carry powerful armament. In consequence, there now is a definite possibility of providing battleworthy air-transportable combat vehicles and thus giving Airborne forces the mobile weapon power which they have lacked so far.

[32] Sir Robert Thompson, *War in Peace: conventional and guerrilla warfare since 1945*, pp. 286-293

[33] For a full description of these systems, see Andrew W. Hull, David R. Markov and Steven J. Zaloga, *Soviet/Russian Armor and Artillery Design Practices: 1945 to Present*, (Darlington, Maryland: Darlington Productions, 1999), pp. 288-308.

[34] For an excellent study of Soviet deep battle, see David M. Glantz, *Soviet Military Operational Art: In Pursuit of Deep Battle*, (London: Frank Cass, 1991).

[35] An excellent study of the six-day war in George W. Gawrych, *Key to the Sinai: The Battles for Abu Ageila in the 1956 and 1967 Arab-Israeli Wars*, (Fort Leavenworth, Kansas: Combat Studies Institute, Research Survey Number 7).

[36] Alla Malakhova, "Russkie na Afirkanskom roge" [*Russians on the Horn of Africa*], Nezavisimoe voennoe obozrenie [Independent military journal], 21-27 November 1997, 4.

[37] Oleg Sarin and Lev Dvoretsky, *Alien Wars: The Soviet Union's Aggressions Against the World, 1919 to 1989*, (Novato, California: Presidio Press, 1996), pp. 133-134.

[38] Viktor, Chugulev, "Bez znakov razlichiya" [*Without any rank or insignia*], Soldat udachi, [Soldier of Fortune], March 1995, 32.

[39] Steven J. Zaloga, *Inside the Blue Berets: A Combat History of Soviet & Russian Airborne Forces, 1930-1995*, (Novato, California: Presidio Press, 1995), p. 182.

[40] Much of the Ethiopian material was previously published in Chapter 9 of *Mysteries of the Cold War*, (Aldershot, England: Ashgate, 1999), edited by Stephen J. Cimbala. Chapter 9 "*Soviet Soldier-Internationalists in Support of Communist Revolutions*" is by LTC Lester W. Grau (R).

[41] Ralph Zumbro, *The Iron Cavalry*, pp. 318-322

The Somalis had come through the Ahmar Mountains by way of the Kara Marda Pass and pushed into the Ogaden as far as the town of Harar, which they were besieging. The Russians were in the habit of giving obsolete weaponry to their Third World clients and [General] Petrov had 70 x ASU-57 assault guns at his disposal. These were little-tracked vehicles of 3.5-tons weight, armed with a 75mm anti-tank gun. They had originally been designed for the USSR's Airborne Divisions but had been superceded by larger weapons.

Petrov also had available 20 Mi-8 and 10 Mi-6 helicopters with Russian pilots. On 5 March 1978, the Cubans air-assaulted the town of Genasene, 17 miles north of Jigjiga, which was a Somali base. Once the town was held, all 70 x ASU-57s were lifted in. Once they organized, they motored south and hit the Somalis in the rear, while the rest of the Cuban force, mounted *in APCs, hit them in a frontal assault. The predictable result was a casualty list of around 3,000 and three days later the Somalis agreed to withdraw from the Ogaden*

[42] Lester W. Grau and Michael A. Gress (eds.), *The Bear Looks Back: A Russian General Staff Retrospective on the War in Afghanistan*, pending publication, chapter 6.

[43] For examples of Soviet Air Assaults, see Lester W. Grau, *The Bear Went Over the Mountain: Soviet Combat Tactics in Afghanistan*, (London: Frank Cass, 1998), chapter 3 and Ali Ahmad Jalili and LTC Lester W. Grau (R), *The Other Side of the Mountain: Mujahideen Tactics in the Soviet-Afghan War*, (Quantico, Virginia: USMC Special Study DM-980701, 1998, chapter 9).

[44] Major Frank Sherman, *Operation Just Cause: the Armor-Infantry Team in the close fight* Cause (Fort Knox, Kentucky, U.S. Army Armor magazine, Sep-Oct 1996) pp. 34-35

[45] CPT Kevin J. Hammond and CPT Frank Sherman, *Sheridans in Panama* (Fort Knox, Kentucky, U.S. Army Armor magazine, Mar-Apr 1990) p.15.

The "C" Company Commander of 3/73d Armor who parachute airdropped in his light tanks to support the 82d Airborne Division writes;

Our first encounter with the Panamanian Defense Force (PDF) occurred as the infantrymen of 1st Battalion, 504th PIR were establishing a supply route from Toucumen International Airport to their initial objective of Tinajitas. The convoy had only moved a few kilometers when it stopped to clear a roadblock located on a bridge. As the *Sheridans* moved to the edge of the highway to support the infantry, SSG Troxell, the lead tank commander, called me on the radio and stated, "This is hell of a place for an obstacle, buildings all around and no cover. It looks like swamps on both sides of the road". As the infantry dismounted and began to execute their obstacle drill, they began receiving automatic weapons fire from the buildings no more than 50 meters away. The lead tank commander opened up with .50 caliber fire as the wing tank commander screamed to his gunner to identify the threat. A moment later, SFC Freeman, 1st Platoon Sergeant, yelled, "I got 'em, concrete building, second floor, fourth window from the right"...He fired a 152mm heat round at the target, ripping through the room, collapsing the right side of the building. The enemy fire stopped and the infantry finished clearing the roadblock

LTC John Barker, U.S. Army, former XO of 3/73d Armor, Letters to the editor section (Fort Knox, Kentucky, U.S. Army Armor magazine, January-February 1997), pg.1

The *Sheridan* with its 152mm main gun was the near-perfect light infantry support vehicle. It could swim. It had thermal sights. It had long-range armor destruction capability equal to or greater than a *Hellfire* missile (check your PH/PK classified data!) The *Shillelagh* with its 152mm HEAT round could blow a hole in a reinforced concrete wall large enough for infantry Soldiers to walk through side by side. An infantry leader could use the external phone, it boasted a flechette' round that could blast 17,000 one-inch nails into enemy infantry as close support, and oh by the way, you could parachute it into combat for those nasty 'forced entry' missions typically laid at the feet of the Paratroopers of the "*Devils in Baggy Pants*", "*Panthers*" and "*Falcons*" of the 82d http://www.geocities.com/Pentagon/Quarters/2116/popguns.htm

[46] Stan Crist, *Too late the XM8?* (Fort Knox, Kentucky, U.S. Army Armor magazine, January-February 1997 pp.16-19

M8 *Buford* or *Ridgway* Armored Gun System: http://www.geocities.com/Pentagon/Quarters/2116/armored.htm

[47] Christopher F. Foss and T. Cullen , *Jane's Armour and Artillery 98-99*, (Coulsdon, UK: Jane's Information Group Ltd., 1998), Armored Personnel Carriers section by country;

India: BMD-2s purchased from Russia
Israel: M113A3-equivalents modified by national industries
Germany: Mak Wiesels 1s and 2s in Airborne units
Russia: BMD-1s/2s/3s in Airborne VDV units
South Africa: El Macho 4x4 ATV
Italy: M113s built under license from U.S.

[48] Mark Bowden, *Blackhawk down!: A story of modern war* (Penguin Putnam books, 1999 and 2000), pp. 95 http://www.amazon.com/exec/obidos/search-handle-form/103-6593991-3811860

Web site based on newspaper series: http://www.philly.com/packages/somalia/ask/ask12.asp

[49] Michael L. Sparks, *Crisis of Confidence in Armor* (Fort Knox, Kentucky, U.S. Army Armor magazine, March-April 1998), pp. 20-44 *http://www.geocities.com/Pentagon/5265/futuretank.htm*

[50] Peter A. Wilson and Charles Gordon IV, *Aero-motorization: pathway to the Army of 2010* (Carlisle, Pennsylvania, U.S. Army War College, Strategic Studies Institute Report, 1998), pp.9 *http://carlisle-www.army.mil/usassi/ssipubs/pubs98/aeromotr/aeromotr.pdf* Army Times November 15, 1993

[51] LTG Francis Tucker, *The Pattern of War* (Washington D.C.: U.S. Government Printing Office 623-279/10260 USMC HQ Washington D.C., 1989), pp. 20-105

Tucker asserts that if war is not properly understood, maneuver ceases and deadlock follows. At that point "siege engines" must be capable of breaking through enemy opposition to regain maneuver. He explains how we live in an *Icarean* world where 3D maneuver from the air will be the dominant war form.

U.S. Army doctrine for urban areas is to have a preponderance of firepower and combat engineering means: *FM 90-10-1 Infantryman's guide to Combat in built-up areas* (Washington D.C.: U.S. Government Printing Office U.S. Army HQ, 1993), pp. 1-55 *http://www.adtdl.army.mil/cgi-bin/atdl.dll/fm/90-10-1/default.htm*

U.S. Army history validates the need for this type mechanical advantage; Robert Black, *Rangers in WWII,* (New York: St. Martin's Mass Market paperbacks, 1994) p. 37: Black, a Ranger veteran said the following about the disastrous Dieppe Raid, where commandos assaulted fortified positions with little fire support other than what they carried in their hands;

"68% of the 4,963 Canadian troops were casualties and 913 were killed outright.... On August 19, 1942, the Canadians did all that flesh could do against fire, but has been proven on battlefield after battlefield against an aroused, entrenched enemy, COURAGE IS NOT A SUBSTITUTE FOR FIRE SUPPORT"

Yet in recent years, firepower and mobility platforms have been retired and not replaced making it more difficult to gain even full 2-D maneuver.

CSM Timothy B. Chadwick, *Death of the Combat Engineer Vehicle* (Fort Leonard-Wood, Missouri, U.S. Army Engineer magazine, December 1996) *http://www.wood.army.mil/ENGRMAG/PB5964/perview.htm*

Alvin and Heidi Toffler, *War and Anti-War: Making Sense of Today's Global Chaos* (New York, Warner Mass Market Paperback, 1995), pp. 10-42 *http://www.amazon.com/exec/obidos/ASIN/0446602590/qid=953747016/sr=1-3/103-6593991-3811860*

Probable cause for the de-emphasis on physical mechanical advantage is the influence of the Toffler's writings elevating digital mental awareness to a stature above the physical plane we still live on and fight for. They see the world as changing by "waves" of human progress:

1st Wave: agrarian, hand to hand combat
2d Wave: industrial age, machine combat
3d Wave: computers, information war

General William E. DePuy USA (ret), *Infantry Combat*, (Columbus, Georgia: U.S. Army Infantry magazine Mar-Apr 1990), pp. 18-23

General Depuy asserts that "*maneuver must be earned*", it is a condition we must fight for and build for and not assume will be available without a fight or constructive effort. *http://www.geocities.com/Pentagon/Quarters/2116/airlandbattle.htm*

[52] Future Combat System web site (Fort Knox, Kentucky, U.S. Army Armor Center and School) *http://knox-www.army.mil/center/dfd/FCV.htm*

Air-Mech-Strike Proposed *Ridgway* Fighting Vehicle

* Based on German *Wiesel*
* C-130 Transportable
* 20 Loaded in B-747
* Para-Drop Capable
* UH-60L Helo-Sling
* Attack= 30 or 40mm Cannon & *Javelin* ATGM
* APC= 7 Troops Total
* Add-on Armor Capable
* Off Shelf *Air*-Mech Option

GEN Matthew B. Ridgway
WW II Airborne/Korean Warrior

Proposed *Ridgway* Fighting Vehicle (RFV)
(Mak and Major Charles Jarnot)

Chapter 4

Air-Mech-Strike: 3-D Army Maneuver Warfare Theory

"The knowledge of the existence of a well trained Airborne Army, capable of moving anywhere in the globe on short notice, available to an international security body such as the United Nations, is our best guarantee of lasting peace. And the nation or nations that control the AIR will control the peace"
—LTG James M. Gavin, Airborne warfare, 1947

"Battles are won by slaughter and maneuver. The greater the General, the more he contributes in maneuver, the less he demands in slaughter"
—Sir Winston Churchill

Maneuver Warfare Synopsis

From the beginning of time until the end of the 20th Century, armies of the world measured maneuver warfare at the foot speed of a Soldier or the "*hoof speed*" of his horse mounted troops. Although frontal assaults were often tried, they rarely produced the decisive results that indirect approach "maneuver warfare" practitioners like Alexander the Great, Genghis Khan or Napoleon achieved. However, the recognized end to "*foot and hoof*" maneuver warfare was seen during the First World War, 1914-1918, caused by the widespread introduction of machines of immense firepower; machine guns, rapid fire artillery and mortars. Strategic maneuver was still possible in 1914 and there were early races of who could get a sizable land force first to a decisive location; but then the affair at the tactical level quickly was reduced to trench warfare of attrition. The loss of operational or tactical maneuver produced some of the bloodiest battles known in human history; where literally millions of men—almost an entire generation—perished for the gain of only a few yards of churned up ground. Out of the First World War carnage came ideas to use machines for MOBILITY—to break out of the trench warfare paradigm using armored track-layers to cross over the trench works of the battlefield. The final offensives of WWI were astounding displays of Allied tanks, infantry, artillery and aircraft working together in a form of primitive mechanized warfare—a lesson that took deep root in the vanquished and tossed aside by the victors. Significant mechanical difficulties born out of the immature mechanized technology of the day quickly set aside the notion of a mechanized return to maneuver warfare in nearly every Army following the Second World War except the Russian and the German armies and to a small band of futurists in the west.[1]

Since WWII several armies, including the U.S. Army, experimented with fielding AFVs that could be transported by aircraft in 3-Dimensions of battlespace to gain positional advantage. While other major armies have fielded Brigade and Divisional size *Air*-Mechanized formations, the U.S. Army has only fielded one AIRBORNE Armor battalion equipped with the M551 *Sheridan* light tank which is now disbanded. The only *air*-transportable combat maneuver forces in the U.S. Army that use vehicles are the anti-tank platoons, companies and cavalry troops found in the light Divisions. All are based on the HMMWV lightweight truck equipped with an armored shell that's proof only against some small-arms fire and shrapnel. These organizations however, are designed to protect foot -mobile infantry Battalions and Brigades at their pace of operations (1-2 mph) and are not intended for independent *Air*-Mechanized maneuver. Aside from some recent studies and analysis from the Army After Next office at Fort Monroe, Virginia, the U.S. Army to date has not yet fielded *Air*-Mechanized 3-D maneuver forces. This is not the case in Europe where several allies and former adversaries have committed sizable resources towards the development of *Air*-Mech capable 3-D maneuver formations. These armies still rely on the dominance of heavy mechanized 2-D maneuver warfare but have increased their force's flexibility through the development of a credible *Air*-Mechanized 3-D capability.

Russian Model

As stated earlier, the Soviet Army and now the Russian Army, has been at the forefront of *Air*-Mechanized development. Before WWII, they experimented with light tanks and armored cars. After the war, starting with the 4-ton ASU-57, for the last forty plus years they have fielded operational Brigade and Division size units with *Air*-Mech capability. Today, the Russian *Air*-Mech model is based on the 8-ton family of tracked AFVs designated the BMD series. The BMD series includes, infantry carriers, direct and indirect 120 mm howitzer-mortars, Air Defense Artillery, engineer and C2 variants. The BMD can be para-dropped or lifted by the Mi-6 or Mi-26 heavy lift helicopters. The Russian Army currently maintains about 1600 BMD AFVs in its Airborne and Air Assault formations.[2] A shortage of lift helicopters and the poor economic resourcing of the Russian armed forces as a whole have greatly reduced its *Air*-Mech capability from the former Soviet days. However, despite these resource challenges the Russian Army continues to push the *Air*-Mech "envelope" as evidenced by their development of the BMD-3 vehicle.

British Model

During the early 1970s, the British Army fielded a light mechanized recon and rapid deployment force of about Brigade strength based on the 8-ton *Scorpion/Scimitar/Spartan* family of tracked AFVs. There are gun, anti-tank, recon, ADA, infantry carrier, mortar and C2 variants. Initially intended for C-130 transportability only, the British Army opted for an 8-ton family of AFVs vice the competing 16-ton design to ensure helo transportability via the RAF CH-47 *Chinooks*. On 12 July 1999, the British 5th *Air*-Mobile Brigade used its *Air*-Mech 3-D capability by sling-loading *Scimitar* AFVs from RAF CH-47 helicopters into their designated areas in Kosovo. A shortage of heavy-lift helicopters, about 50 total in the entire British inventory, limits the British *Air*-Mech model to a Brigade-size force with a single-Battalion-at-a-time lift capability.[3]

German Model

Following a close examination of both the Russian and British models, the German Army opted to incorporate *Air*-Mechanization into its Airborne forces. In 1992, the German Army completed fielding some 400 tracked 3-4 ton *Wiesel* AFVs into their Airborne Brigade. By going a size lighter than the Russian or British *Air*-Mech designs, the German model gained the greatest air-transportability possible making its *Air*-Mech forces sling-loadable by the smaller but much more numerous UH-60 and *Super Puma* NATO helicopter fleets. In addition, the German *Air*-Mech force can be readily transported to another theater using commercial large body jets. About 25 *Wiesels* can be loaded in a single standard Boeing 747 converted to cargo! The *Wiesel* comes in recon, gun, ADA, troop carrier, and C2 variants. The German Army utilizes their *Wiesel* forces for Cavalry-like operations involving reconnaissance and anti-tank missions. Like the British, a shortage of heavy lift helicopters, about 100 total in the entire German inventory, limits their *Air*-Mech model to a Brigade-size force with a dual-Battalion-at-a-time lift capability.[4]

U.S. Army Today

The evolutionary adherence to 2-D heavy mechanized maneuver doctrine has led to the U.S. Army currently fielding the world's heaviest example with the 70-ton M1 *Abrams* tank and 33-ton M2 *Bradley* Infantry Fighting Vehicle combination.[5] Most of the firepower in the American Army is centered in the Heavy Divisions with their 600 tanks and Infantry Fighting Vehicles. Both weapon systems are designed to fight and survive in the crucible of massed direct-fire engagements with other mechanized 2-D forces. The excessive weight however, makes such forces extremely difficult to deploy by air and very slow to move by sea. Their weight results in poor terrain agility often limiting their employment to valley floors and road networks and severely taxes a nation's road and bridge network. For example, in 1995, it took three days for an U.S. Army Heavy Brigade to cross the Sava River in Bosnia…unopposed![6] The U.S. Light Divisions are the opposite with literally all infantry formations being foot mobile at 1-2 mph at best, requiring extensive support from Corps level transportation units just to move from one location to another. They lack tactical mobility, have no armored protection and employ mainly man-portable firepower and light towed artillery. They are therefore relatively easy to deploy but have few capabilities against anything other than a dismounted insurgency or foot mobile civil riots. Only the Air Assault Division has significant numbers of heavy attack helicopters to engage enemy armored formations.

Where Do We Go From Here?

The difficulties experienced in trying to deploy ground forces in the Balkans and other contingencies around the globe has provided the setting for the call to rethink the U.S. Army's force structure. How does an Army achieve the goals stated by the CSA using the well-established doctrine of mechanized maneuver warfare? Lighter armored vehicles can be fielded which are easier to deploy from bases to the battlefield, but offer little enhancement in terrain agility if they are not helicopter transportable and capable of aggressive cross-country movement. Given the maneuver speeds and limitations of a 2-D mechanized warfare, the argument can be made that lighter Armored Fighting Vehicles (AFV) will only be more vulnerable then their heavy armored predecessors unless they can be flown by helicopters and traverse terrain heavier AFVs cannot cross. If the American Army is going to lighter AFVs, then why not go light enough to achieve helo-transportability and dramatically out maneuver your heavy opponent with true 3-Dimensional maneuver warfare? The difference in protection from light AFVs being considered by the U.S. Army's experimental interim Brigade test beds in Fort Lewis Washington, are not any meaningful improvement over some European helo-transportable designs. The *Air*-Mech-Strike concept proposes to change the current paradigm of a 2-D maneuver-only Army to a fully integrated 3-D force.

Air-Mechanization Defined further

The central idea behind *Air*-Mechanization is to design a land combat force capable of *air*, mechanized and dismounted maneuver in order to achieve decisive action through positional advantage. By its very nature, lightweight *Air*-Mechanized forces can rapidly deploy strategically, possess above-average 2-D terrain mechanized agility and are relatively easy to sustain. *Air*-Mechanized forces can quickly gain positional advantage by being 3-D air-inserted from either fixed or rotary-wing aircraft. Once landed, these forces can quickly transition to mechanized maneuver with light armor for protection from small arms and shrapnel, possess great lethality through lightweight high tech weapons and prevent deadly surprise meeting engagements with enemy heavy armor through digitized situational awareness.

Moves Towards *Air*-Mechanization

Technological challenges in helicopter lift capability and *Air*-Mech vehicle curb weights require that the transition toward *Air*-Mechanization be gradual with initially combined organizations of *Air*-Mech capable formations teamed with scaled down heavy task forces. The tank and other armored fighting vehicles are not going away under this concept but are re-invented as lightweight high tech "*Aegis Cruisers*" vice their current "*Iowa Battleship*" design. Modern technology is rapidly making the fielding of 70-ton heavily armored vehicles un-necessary. Advances in weapons and sensors are making the direct-fire crucible between massed formations of heavy armor at close ranges a very unlikely scenario for the 21st century. The surprise close-in direct fire engagement resulting from maneuver through compartmented and vegetated/urbanized terrain still requires the armored gunfighting tank. The urbanization of the world demands high energy weapon systems to destroy tanks, bunkers and buildings. New technology however, demonstrates that this can be done with guided lightweight hyper-velocity rockets and slower high-explosive missiles. The need to mount heavy high-velocity cannons is rapidly becoming obsolete. With the need diminishing to employ massive armor plate on weapons systems comes new possibilities for air-maneuver. In the *Air-Mech-Strike* concept battlefield digitalization is fully exploited to shift the primary defeat mechanism from direct line-of-sight engagements to over-the-horizon Precision Munitions Attacks (PMAs). The Army must learn the Navy's lesson that you can't build a ship with enough armor to withstand a precision munitions attack from over-the-horizon. The best defense against such a menace is to quickly gain positional advantage to see and then kill the enemy platform first.

Digital Battlefield & Precision Munitions Component

The liability of thin armor, necessary to gain air transportability is mitigated somewhat by the employment of high technology lightweight weapons with great over-the-horizon stand-off ranges i.e. Enhanced Fiber Optic Guided Missiles (EFOGM with 15 kilometers range and vertical hit lethality). Even direct-fire weapon technology now provides kinetic energy penetrators equal to heavy main tank gun munitions at weights less than 200 pounds such as the hyper velocity rocket powered Line of Sight Anti-Tank missile (LOSAT, with 5 kilometer range and one-mile-per second velocity). The recent development of battlefield digitalization through remote air and ground based sensors provides the light armored force with unprecedented "situational awareness" giving Battalion and Company Commanders 3 times the LOS (Line-Of-Sight) or greater battlespace awareness. This in turn, provides the means for targeting long-range precision weapons and reducing the risk of surprise meeting engagements with heavy enemy armor. Finally high technology rockets and guided missiles provide long range precision fire support with UH-60 air transportability via trailer-mounted Multiple Launch Rocket Systems and Army TACtical Missiles. Without a robust digitalization component, an air inserted lightweight armored force could be at extreme risk in a surprise meeting engagement with heavy enemy armor. Battlefield digitization efforts, well underway in the Army's Force XXI project, would be better spent on a faster moving *Air*-Mech force and decisively maneuver inside an enemy's decision cycle than a 2-D only force.

Vehicle Component

Armored Fighting Vehicle technology has advanced steadily since 1939 where today high-tech suspensions, composite armor and high power engines are driving the weight versus capability downward.[7] Several nations have developed lightweight AFVs to enhance air-transportability while still retaining the near capabilities of the heavier predecessor. Such an example can be found in a comparison between the 15-ton Russian BMP Infantry Fighting Vehicle and its lighter cousin the 8-ton BMD. Both have the same weapon systems with only a modest advantage in armored protection for the former.[8] As the weight versus capability of armored vehicle decreases, the lift capacity of helicopters is increasing. The *Air-Mech-Strike* concept is largely based on the claim that the intersection of these two curves, pushed upward even further by digitalization, makes a supplemental 3-D maneuver force possible today. And if trends continue with higher intersection points, eventually combined 2-D and 3-D *Air-Mech-Strike* will displace heavy 2-D only mechanized maneuver as the dominate warfighting doctrine for the remainder of the 21st Century.

Aviation Component

Recent developments in both rotary wing and Short Take Off and Landing (STOL) fixed wing aircraft lift capability has greatly increased the payload of vertically-inserted combat forces. This in turn, has expanded weapon system options that have correspondingly enhanced ground mobility, lethality and survivability. Ranges have likewise increased, which have expanded the strategic, operational and tactical power-projection capabilities of an air-inserted combat force. The increased range and speed of new platforms also greatly reduces the need to move vast amounts of fuel stocks forward to support large scale helo operations by increasing the strike distance from Theater Support Bases (TSBs). However, the most significant aviation factor for the American Army to consider in deciding whether or not to adopt the *Air-Mech-Strike* concept is that the U.S. Army is the single largest operator of lift helicopter on the planet. While other foreign *Air*-Mech models may have just enough lift helos total in their national inventories to lift a battalion or two at a time, the U.S. Army's current 1200 UH-60 *Blackhawk* and 430 CH-47 *Chinook* fleets could lift the equivalent of two Divisions![9]

The Cost Component

For the U.S. Army, the adoption of an *Air*-Mechanized model for the new proposed Brigades will cost less than other competing 2-Dimensional force designs. The savings begins with the substantially lower vehicle costs, a German *Wiesel-2* for example is only half the cost of a LAV-25. The M113A3 *Gavin*'s maintenance and fuel costs are equal to a large armored car; the small *Ridgway* is substantially lower and its fuel economy is 3 times better than its larger counterparts which means lower cost sustainment requirements. The greatest savings however, are in the tremendous reduction in the cost of strategic deployability. A typical German model *Air*-Mech Brigade of 300 vehicles only requires 15 X Boeing 747s to transport, much fewer sorties than transporting a LAV Brigade and no need for scarce Military Cargo aircraft. The standard U.S. Army UH-60L *Blackhawk* helicopter can also be transported in cargo 747s. The very expensive to obtain, large helicopter force required to effect *Air*-Mechanized maneuver **already exists** in the U.S. Army. The increase in rotary-wing fuel and maintenance cost of an *Air*-Mech force is more than compensated by the great reduction in large cargo fixed-wing fuel and maintenance costs associated with deploying a 2-Dimensional all-LAV-25 type force or worse an all 70-ton M-1 tank based force. Finally, the high-tech sensors and precision weapons cost would be the same for both a 2-Dimensional heavy armored force and a lightweight 3-D *Air*-Mechanized force.

The Proposed American *Air*-Mech-Strike Model

The advantage at being a "late comer" to the concept affords the American Army an opportunity to evaluate the different models, compare the capabilities and examine how they might be adopted to the current U.S. Army force structure. The two great resources found in the current Army is a very robust helicopter lift force and an excellent battlefield digitization base of technology developed under the Force XXI initiative. Other chapters in our book explore in detail the possible lay out of an American Army transitioning towards *Air*-Mechanization. However, for the purpose of explaining the mechanics of the *Air*-Mech-Strike concept we will examine in rough terms a possible *Air*-Mech model utilizing the current UH-60 *Blackhawk* and CH-47 *Chinook* fleets and the German *Wiesel*. We'll examine a Brigade size force executing a notional scenario in the year 2005 and draw a technology road map on a possible path to full *Air*-Mechanization by the year 2020 with the Future Transport Rotorcraft (FTR) and Future Combat System (FCS) force structure. Other tracked and wheeled 3-4 ton AFVs exist that could be adopted in lieu of the *Wiesel* to achieve similar results. The German *Wiesel* however, represents a large fleet of combat proven vehicles fielded in Brigade strength. The German Army, a major NATO ally of the United States, has operated the vehicles for 11 plus years. As such it represents a large data base of actual field and combat experience over other "prototypes" offering only theoretical data or estimates of expected performance.

German *Wiesel* Adopted as the "*Ridgway* Fighting Vehicle"

Our model adopts the base German *Wiesel* vehicle and makes some modifications. We call the Americanized *Wiesel* the *Ridgway* Fighting Vehicle (RFV). The RFV comes in the two base models as the original German *Wiesel*, a 3-ton Reconnaissance and Security (R& S) platform and a 4-ton carrier (PC) version. A key modification that we believe would be necessary but could be adopted without great expense or refit is the top part of the hull being constructed out of Kevlar forming a removable protective tub to facilitate split loading. The Kevlar or like composite armor would offer a weight savings over the current use of steel in the German *Wiesel*. The RFV would therefore be a little lighter than the German version and allow for split loading, a technique that would facilitate helo transportability in high elevations and hot temperatures.

The RFV-RS, with a crew of two, would also incorporate a U.S. MK-19 40mm Grenade Machine Gun initially then upgrade to a M-230 (ASP-30) 30mm autocannon (same as on the AH-64 *Apache*) vice the current German 20 mm and have a pair of signature-less, *Javelin* top-attack fire/forget anti-tank guided missiles launchers capable of engaging/destroying enemy armored vehicles, helicopters and building/bunkers. The *Javelin* Forward Looking InfaRed (FLIR) sight would be adopted for both missile and cannon use. The use of *Kevlar* armor would provide the weight savings to keep the vehicle combat loading at 6000 pounds and the low-velocity, low-recoil 30mm would not require additional recoil dampening. The RFV-RS would also incorporate a lightweight combat laptop computer with GPS and a laser designator. The net result over the German *Wiesel*-1 is a recon platform with significant increase in firepower and anti-tank capability along with better day and night optics and better battlefield situational awareness.

The RFV-PC carrier, 8000 pounds, would likewise have a removable top *Kevlar* armored modular shell to allow for split loading and have greater internal room to facilitate the carrying of a four-man infantry fire team and a crew of two. The German 7.62 mm medium machine gun mounted in the vehicle commander's cupola would be replaced initially by a Mk-19 40 mm automatic grenade launcher and a FLIR incorporated for day/night optics along with a combat laptop, Global Positioning System and a laser designator. Other versions like the German 120 mm Mortar Carrier and Air Defense models would be adopted as is with the only modification being the use of *Kevlar* removable shells. Both the RFV-RS and the PC versions would have provisions for add-on armor plates that could be mounted by hand and delivered in follow-up aircraft to enhance the vehicle's vulnerability to heavier weapons and Rocket Propelled Grenades (RPGs).

M113A3s Adopted as the "*Gavin* Fighting Vehicle"

One infantry Battalion in each Brigade would be equipped with a modernized M113A3 dubbed a "*Gavin* Fighting Vehicle". This type-classified in wide Army-service vehicle would constitute the bulk of Air-Mech capable forces by USAF fixed-wing aircraft airland/paradrop with U.S. Army CH-47F helicopter slingload/streamlined external loading as possibilities in the near future.

All Terrain Vehicles

The explosive technology in the commercial market for lightweight 4 and 6- wheel All Terrain Vehicles (ATV's) has provided the U.S. Army with an excellent helo-transportable reconnaissance platform. Two mid-size 4 X 4 ATV's can be loaded internally in a UH-60 *Blackhawk* and with auxiliary fuel tanks, and inserted via a winch through the vegetation up to 500 kilometers radius and return without refueling. The internal load means closer to terrain masking contours and thus more survivable flight modes as well. Two ATVs could be mounted on a plank running across the floor on the left and right side of the UH-60. A 6-man recon team could be inserted with their large size ATV's sling loaded by a UH-60 out to 350 kilometers radius. ATVs can be made to run fairly quite and with a 100 to 200 pound cargo capacity in addition to a 250-pound Soldier, could carry sophisticated infrared sensors and night sights, robust communication gear, laser designators, a dozen anti-tank mines and even the long range "fire & forget" *Javelin* anti-tank weapon with two missiles. Ultralight ATVs weighing as little as 125 pounds can be rigged to be driven remotely via cameras and a transmitter or payed-out fiber optic line. These Un-manned Ground Vehicles (UGV) can be inserted deep by helicopter or winched down through the trees contributing to the excellent situational awareness of the ground commander. ATVs are inexpensive, easy to maintain and offer excellent ground mobility with superb helo-transportability.

AIR-MECH MATH

Helicopter Lift Factors

The principle factors that determine helicopter lift capacity are the aircraft's transmission and rotor system, the power produced by the engines, the turbine temperature, fatigue load suffered by key components like drive shafts, gears and bearing surfaces and the outside air density. Extensive testing is done to construct performance envelopes where the aircraft can be operated safely. A given helicopter is then "allowed" certain payloads based on configuration, air density (measured by temperature and elevation above sea level) and power available. So just because one might read in a Defense industry-type publication that a helicopter's hook might be rated to sling a certain weight, that does not normally mean how much it can lift! Quite often the actual lifting capacity is much lower.

Maximum Payload

The Maximum payload for a helicopter normally means the most weight that can be lifted under the most ideal conditions. That means an aircraft with very little fuel on board and just the minimum equipment necessary for flight. It also assumes near sea level elevations and low outside air temperatures. For military helicopters such extreme "stripped" down *Air*-Mech Configurations to achieve max lift may be necessary when picking up heavy loads for short flights across an obstacle like a river or retrieving a crashed aircraft in a remote area.

Practical Payload

The load that a helicopter can lift under most conditions is referred to among many pilots as the "practical" payload. This normally means a combat equipped aircraft with additional survival gear, self-protection machine guns and a full tank of internal fuel. Temperatures are generally accepted as being cool at 75-80 degrees F and elevation at or below 2000 feet. In the *Air*-Mech-Strike design we emphasize practical payloads.

High-Hot Payload

This term normally is used to describe what the aircraft is capable to lift in conditions where the elevation is at 4000 feet or above and temperatures at or above 95 degrees F. The exact figures of 4000/95 is known officially by the Department of the Army as the "*High-Hot*" day. Most helicopter performance at or above this region begins to deteriorate rapidly. The thinner air caused by the higher elevation and or hotter outside air temperature causes the turbine engines to begin to overheat. This causes the pilot to limit his demands for power and thereby reduces lift capacity. Its important to note that this degradation is not linear! A helicopter's lift performances falls off sharply after the High Hot conditions are reached. Therefore it may be that a given helicopter has the horsepower to lift a certain load but may be limited by turbine temperature. It only takes a few seconds of over temperature to destroy a turbine engine and the effects of hot operation is often accumulative.

Combat Radius

This term refers to the approximate distance an aircraft can fly out to, perform its mission and return without refueling enroute. A model profile is used to set conditions that would normally be expected such as the need to always maintain a 20 minute fuel reserve, typical sling load speeds etc. The combat radius figure is a good planning tool but actual distances can only be determined by using current weather and aircraft information. The profile of the mission may greatly influence this figure as well. For example, the combat radius of a reconnaissance helicopter may be very short compared to a lift aircraft because the reconnaissance mission profile may require at least 45 minutes of station time at the objective. A lift aircraft typically spends less that 3 minutes dropping off a combat load.

Exceeding Normal Operating Parameters

During extraordinary circumstances an aircraft can exceed its limitations such as during an actual combat mission or during an emergency. During *Desert Storm* for example, several Army aircraft were given authorization to carry heavier than normal loads. The danger in doing this is that the built in safety margins are lowered and the service life of major components like engines and transmissions can be reduced. Some of these effects may take a few years to manifest themselves like cracks in transmission gears. These accumulative effects of over stressing an airframe have been suspected in a few fatal crashes over the years. The *Air*-Mech-Strike concept uses UH-60 and CH-47 aircraft to the limits of their lift potential but we are careful to remain well within current operator's manual limits and judiciously apply increases in data attributed to future designs not yet in production like the CH-47F model. *Air*-Mech-Strike is not contingent on Desert Storm like performance limitation waivers.

UH-60 *Blackhawk Air*-Mech Math

The UH-60 Alpha model *Blackhawk* has a practical payload of about 6,000 pounds. This allows the aircraft to sling a *Ridgway* Fighting Vehicle Reconnaissance and Security version with its two-man crew carried internally out to a combat radius of about 150 kilometers. A UH-60A could not practically sling a RFV Personnel Carrier version which weighs 8,000 pounds. Using RFV *Air*-Mech Configuration 1; removing the top kevlar armored tub which would weigh about 2000 pounds, this split loading would allow UH-60As to sling the RFV-PC employing additional aircraft to carry the troops and top tub. With the tub removed, the RFV is in *"Ridgway* Air-Mech-Configuration 1" or *Ridgway AMC 1*. The RFV with tub and any appliqué armor would be for situations where the aircraft capacity situation doesn't require split-lifting and would be referred as *Ridgway Air*-Mech-Configuration 2 or *Ridgway AMC 2*. These situations might be when RFVs are parachute air-dropped (para-mech) from USAF fixed-wing aircraft or transported by larger CH-47D/F helicopters for 3-D maneuver.

The UH-60 Lima model *Blackhawk* has a practical payload of about 8,000 pounds.[10] This allows the aircraft to sling a RFV-RS with its two man crew carried internally and an additional internal *Robertson* auxiliary fuel tank out to a combat radius of about 250 kilometers. An UH-60L could practically sling a RFV-PC with six troops internally loaded out to a combat radius of 150 kilometers.

the needed level of lift required to effect sizable *Air*-Mech maneuver, most of the Army's aircraft should be to be organized under Corps Aviation Command (COAVCOM) and assigned one each per Corps. Each ground maneuver Division would have a single Aviation Squadron of two Troops of 8-12 UH-60s each and a single troop of 8 *Kiowa Warrior* recon aircraft along with an ATV equipped ground recon troop. This Aviation Squadron could perform small scale anti-armor missions with UH-60s firing *Hellfire* and *Kiowa Warriors* lazing targets, Air Assault recon squads and platoons from the ground ATV troop and conduct command and control aviation missions.

The COAVCOM would be organized into three Brigades with an attack, lift and support regiment each. Each Brigade would run a four-month training cycle so that one would always be ready to rapidly deploy to support a designated ground maneuver Brigade or Division. In this way all ground Divisions could expect massed lift and attack support to effect *Air*-Mech maneuver if they are deployed. Active, National Guard and Reserve aviation units would be incorporated into the COAVCOM. Each Aviation Brigade and/or Regiment would have National Guard and Reserve staff members to supplement the predominately active headquarters. National Guard and Reserve Battalions would be incorporated into the Regiments along side active units.

To facilitate aircraft maintenance and fleet service life cycles, each COAVCOM would operate a series of geographically located aviation training facilities with small fleets of attack and lift aircraft. Battalions would draw aircraft for training much in the same manner as reserve units do at Mobilization And Equipment Sites (MATES). Within each Regiment there would be a deployable maintenance Battalion modeled after the USAF wing organization. This would free up Aviation Battalion Commanders to concentrate their training effort on tactical flight training and not on hangar maintenance operations which is the current case in most aviation units today. The MATES system of aviation training means that only a third of the total fleet would be used for flight training with a very favorable monthly hour utilization rate that would facilitate maintenance and fleet service life. A third of the fleet could then be pre-positioned in flyable storage overseas to support rapid power-projection. The remaining third of the fleet would be used for school house training, equipment upgrades, airframe overhauls and system testing. The result would be that Army Aviation Battalions could mass to meet mission requirements and more rapidly deploy into theater, operate low time aircraft in a surge operation and have lower operating costs via the efficiencies gained from aviation MATES sites.

The final *Air*-Mech-Strike aviation requirement is the need to maximize aircraft sortie rate per day. Currently units have authorized only one crew per aircraft. We propose that there be at least two crews authorized per aircraft in the unit table of organization and equipment (TO & E). This can effectively double the daily sortie rate of an aviation Battalion and facilitate 24 hour operations. Additional aircrews are easier to deploy than addition airframes. With most maintenance requirements handled by the Regimental maintenance Battalion, aviation units will be able to concentrate on getting all aircrews fully mission ready. The cost of the additional aircrews could be mitigated in large part by an aggressive program to retain aviation crew members and even employ individual reservists in active battalions as the co-pilots. Aviation training should be modified more along commercial aviation training lines where experienced aircrew members can receive in-house transitions without the need for costly trips to Fort Rucker for training courses. All graduates from the Army Rotary-wing school should be aggressively recruited to stay involved in flight duties both when on active duty and managed closely in flexible reserve flight training pools. Today many trained flight crews leave active service and are not retained in reserve aviation units due to the difficulties with finding a funded flight position nearby. The aviation MATES system of training will greatly enhance the retention of flight crews in the reserve components.

Air-Mech-Strike, "Do-able", Affordable & Rapidly Deployable

The capabilities of the European Air-Mech models serve as the base line data for our proposed American Air-Mech-Strike model. Arguments can be raised concerning this or that performance claim but the weight of evidence is overwhelming that the ability to execute 3-Dimensional maneuver is already here.[12] The U.S. Army can spend its limited resources on fielding five 2-Dimensional police-like armored car Brigades with limited Air-Mech capabilities only by USAF fixed-wing aircraft or for less cost field a fully 3-Dimensional Air-Mech capable force through-out the entire Army. Delay in adopting Air-Mechanized maneuver and we may be taught the next phase in Air-Mechanization not by our British and German allies but perhaps by a resurgent Communist China as they force their will on Taiwan via a Soviet Style Air-Mech attack across the 100 mile stretch of water. As the late Brigadier General Simpkin stated, "The Race is to the Swift". [13]

---

[1] John Keegan, *Second World War* (New York: Viking Penguin Books, 1989), 437 and 438.

Kenneth Macksey, *Guderian: Panzer General* (London: Greenhill Books Inc., 1992), 57-97.

FM 100-5, *Operations* (Washington, DC: GPO, 1993), 6-14.

Major Emmett Shaffer, member of 160th Special Operations Helicopter Regiment planning staff, interview by Major Jarnot, 12 November 1995.

[2] U.S. Army Field Manual (FM) 100-2-3, *The Soviet Army: Troops, Organization and Equipment* (Washington, DC: U.S. Government Printing Office [GPO], 1991), 4-144 and 5-37.

[3] Richard E. Simpkin, *Race To The Swift, Thoughts on Twenty-First Century Warfare* (London: Brassey's Defence Publishers Ltd., 1987), 119.

[4] 8 Bundeswehr, organizational document received from the German liaison office (Fort Leavenworth, Kansas: USACGSC, February 1996). *http://www.geocities.com/Pentagon/Quarters/5265/ozelotssaveday.htm*

[5] *Jane's Armour and Artillery 95-96,* edited by Christopher F. Foss and T. Cullen (Coulsdon, UK: Jane's Information Group Ltd., 1996), 137 and 142. A review of tanks and infantry fighting vehicles reveals that the U.S. M1 *Abrams* tank and M2 *Bradley* Fighting Vehicle are the heaviest vehicles fielded in significant numbers by a major military power.

[6] CNN report on crossing the Sava River in Bosnia, 12 January 1995.

[7] Kenneth Macksey, *Guderian: Panzer General* (London: Greenhill Books Inc., 1992), 57-97.

COL Douglas A. Macgregor, *Breaking the Phalanx*: A New Design for Land power in the 21st Century, Westport, CT: Praeger, 1997.

[8] U.S. Army ST 100-3, *Battle Book* (Fort Leavenworth, KS: USACGSC, 1995), 3-1 and 3-9. The total numbers of key weapon systems counted for each Division model compared with the lift capacity of the C-17 *Globemaster III* are found in *Jane's All the World's Aircraft 1996*, edited by P. Jackson (Couldsdon, UK: Jane's Information Group Ltd., 1996), 584-88.

[9] U.S. Army Student Text (ST) 100-3, *Battle Book* (Fort Leavenworth, KS: U.S. Army Command and General Staff College [USACGSC], 1995), 3-1 and 3-9.

This is based on Major Jarnot's experience as an Observer/Controller (OC) for 28 rotations at the National Training Center, Fort Irwin, California, from 1989 to 1990.

International Institute for Strategic Studies, *The Military Balance 1995-1996* (London: Institute for Strategic Studies, 1996), p. 45 and 115.

Dale R. Steinhauer, resident expert on Middle East wars, interview by Major Jarnot, 6 April 1996, Army Knowledge Network, U.S. Army Command and General Staff College, Fort Leavenworth, Kansas, 6 April 1996.

This comes from Major Jarnot's personal experience in training with the AH-64, AH-1 and UH-1 helicopters at Fort Rucker, Alabama, while assigned to the Aviation Training Brigade from 1993 to 1995.

Chief Warrant Officer 4 Ronald Ferrell, U.S. Army Aviation Center resident expert on RAH-66 capabilities, interview by Major Jarnot, Fort Rucker, Alabama, 6 April 1996.

[10] Jane's Armour and Artillery 95-96, 675. The weight of the rocket pod is 5,600 pounds, allowing for a 2,400-pound trailer. Total weight=UH-60L external limit, 8,000 pounds.

Ibid., p. 675, p. 364 and p. 365.

Jane's All the World's Aircraft 1996, p. 497.

[11] CNN footage showing Israeli AH-1 and AH-64s firing at *Hezbollah* positions in daylight, 19 April 1996.

[12] BG Huba Wass de Czege, lecture on advanced warfighting in the early 21st century (Fort Leavenworth, Kansas: U.S. Army Command and General Staff College, 10 January 1996).

[13] Ibid., p. 123.

Wallace P. Franz, "*Airmechanization: The Next Generation*," *Military Review* (February 1992), 59-64.

Air-Mech-Strike Proposed *Gavin* Fighting Vehicle (GFV)

WWII and post-war Airborne General James M. Gavin

* Based on the M113A3
* C-130 Transportable +troops
* C-17: 5 per plus troops
* Para-Drop Capable
* CH-47F Helo-Slingable
* 30mm-40mm Auto Cannon
* FLIR camouflage strips
* Band-tracks give high road speed, fuel economy, stealth

* *Javelin* fire/forget ATGM
* ACAV-Like Gun Shields
* Add-on Armor RPG/30mm
* Out-Performs wheeled armored cars
* Low Cost Option $150K to convert M113A2 to GFV

Proposed AMS *Gavin* Fighting Vehicle (GFV)
(U.S. Army and Major Charles Jarnot)

UH-60-AM (Air-Mech) BLACKHAWK
STREAMLINED EXTERNAL LOAD (SEL)

- Wide Chamber Blades (WCB) & Dash-8 T-701 Engines
- Higher Payload = Internal Robinson Aux Fuel Tank & More Range
- SEL Modification Retracts to Lima Profile
- Closer Tail Wheel Allows Ship Board Landing
- Wiesel-2 APC + 7 Troops 400 Km Radius

★ 100+ Knot Cruise vice 80 Knot Slingload speed = More Range/per gal
★ Rolling Takeoff = 25k+ Max GWT with 10,000 Pound Payload (+25%)
★ More Maneuverability & Closer Terrain Flight = Greater Survivability

UH-60 Steamlined External Load (SEL) of RFV (John Richards)

ROTARY-WING AIRCRAFT DELIVERY METHODS

V/TOL AIRLANDING   **HOVERING SLING-LOADING**

STREAMLINED-EXTERNAL LOADING (SEL) AND V/TOL AIRLANDING

Rotary-wing Air Delivery methods
(U.S. Army Armor magazine and Mike Sparks)

FIXED-WING AIRCRAFT DELIVERY METHODS

Low-Velocity Parachute Airdrop (LVAD) of personnel/vehicles/supplies

STOL AIRLAND

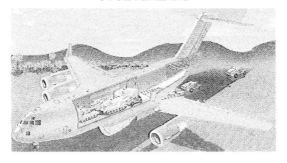

Airdrop: Low-Velocity parachutes (Para-Mech)
(U.S. Army and Mike Sparks)

Chapter 5

The 12 *Air*-Mech-Strike Axioms

"The Paratrooper motto, 'Strike and Hold', indicates basic tactical immobility. With the perfection of air vehicles that can land and take-off almost vertically, using unprepared ground sites, the tactic has changed from 'strike and hold' to 'strike and strike and strike', allowing a continuous series of Air-mobile thrusts."
General John R. Galvin (USA, Retired), *Air Assault: The development of Airmobile Warfare*

The viability of the *Air*-Mech-Strike concept depends on a set of tactical and organizational principles we term "the 12 Axioms" of *Air*-Mech-Strike operations. Most of these rules already exist as Standard Operating Procedures (SOPs) in various military organizations with focused tasks such as found in special operations units. However, the 3-Dimensional nature of *Air*-Mechanized maneuver requires these select "SOPs" to be closely adhered to enhance the success of the concept. The actual military situation or current resources available may deem it necessary to deviate from them but the *Air*-Mech-Strike Commander and force builder must understand and manage the risk that comes with doing so. The German Army of the 1930s developed similar principles and techniques that supported the concept of mechanized maneuver. As an example, the German mechanized force builders deemed it essential that each tank have a radio to facilitate close coordination on the move to maintain the momentum of mechanized maneuver. Likewise, *Air*-Mech-Strike maneuver requires certain tactics, techniques, procedures and some specialized equipment to gain the desired outcome of 3-Dimensional maneuver. The 12 Axioms as a group is another unique aspect of the *Air*-Mech-Strike model versus the British, German or Russian *Air*-Mechanized concepts. Their models contain some but not all of the stated features found in the axiom list.

1. Maximize Displaced Drop, Landing and Pickup Zones: Drop, land and pick-up friendly forces where the enemy ain't! *Air*-Mech-Strike positional advantage does not depend on landing troops near or on top of the objectives where enemy ADA and other direct-fire weapons systems are expected to be thickest. Use displaced DZs, LZs and PZs and then motor to the decisive areas. The fact that an *Air-Mech-Strike* force can attack an enemy from 360 degrees is sufficient to present the opportunity for decisive positional advantage, it does not have to be air-inserted at the heart of his defenses.

The mechanized capability of *Air*-Mech-Strike adds a crucial dimension to the maneuver commander's decision making process concerning landing area selection. The current U.S. Army Airborne/Air Assault doctrine depends on relatively open fields to land Paratroops or helicopters close to objectives so the dismounted troops can reach their targets in a reasonable amount of time at foot speeds. *Air*-Mech-Strike Commander can utilize a Circle Of Insertion (COI) that is probably 10 times larger or more than the requirements of a dismounted assault. The Streamlined External Load (SEL) capability of *Air*-Mech capable UH-60s and CH-47s also allows Air Assault forces to be inserted in complex or urbanized terrain as opposed to open fields. Guided Parachute systems will also facilitate the insertion of Airborne forces and their fighting vehicles into smaller terrain openings from long-range stand-off, keeping delivery aircraft from actually landing/exposing themselves in the Landing/Assault Zone.

2. Avoid Maneuver Mode Mismatches: Avoid missmatches in maneuver where air insertion is used when mechanized maneuver would have been more effective or where dismounted action would have been better than mechanized maneuver. The challenge of a fully integrated 3-D force is to know when and where to switch modes of maneuver.

The significant displacement of landing and pick-up zones along with the ability to use complex and urban terrain for insertion and extraction greatly enhances the survivability of AMS maneuver forces. These advantages however, complicate the Commander's decision on when and where to switch from air to mechanized to dismounted maneuver. This decision is similar to the one made by mechanized infantry Commanders for the last 60 years, that being when and where to dismount in a tactical situation. If during the intelligence preparation of the operation, the enemy air defenses on the given terrain impose an unacceptable risk, then the landing zones need to be displaced further away or mechanized maneuver needs to be employed. If anti-tank defenses pose unacceptable risk then dismounted maneuver needs to be employed.

Unless a maneuver force is facing a "*Maginot*" line of fortress defenses, there will most likely be areas within the desired COI where enemy air defenses are weak or can be adequately suppressed and or destroyed. Most land forces today cannot afford to put a shoulder-fired SAM behind every tree! If an *Air*-Mech force is facing a "*Maginot*" line of fortress like defenses such as found on the border between North and South Korea, short-range tactical *Air*-Mech-Strike assaults may not be possible within such a well-developed defense structure. In such situations *Air*-Mech-Strike forces should be employed in a more operational role and in concert with joint force support conduct longer range *Air*-Mech-Strikes to outflank the entire fortress network, much in the way that German mechanized forces outflanked the actual "*Maginot*" line in 1939.

3. Over-The-Horizon Situational Awareness: The need to minimize vehicle weight in order to achieve helo-transportability requires acute attention to increasing situational awareness beyond the horizon of line-of-sight to prevent surprise meeting engagements with enemy heavy armored formations or stumbling across prepared defenses. Battlespace situational awareness has always been sought by the larger force Commander however, the speed at which *Air*-Mech-Strike maneuver operates at and the relatively thin armored vehicles associated require this level of awareness down to the Company and Platoon level. The increased need for detailed situational awareness of lighter *Air*-Mech forces is reflected in our force structure recommendations where we built more reconnaissance capability in the Brigade force structure models.

For the Generals and Colonels it is currently satisfactory to know from data down-links or other remote intelligence that an enemy armored force is moving toward his forces. Such information will be posted and tracked every five minutes or so as a red unit symbol on a map at a command post. But in *Air*-Mech-Strike operations, it is crucial to the Captains, Lieutenants and Sergeants that they know where the next "T-80" will come at them in the next 60 seconds. In order to have that level of beyond tree-line situational awareness, Companies and Platoons will have to have their own mini-UAVs and UGVs that are about the size of a football or a wagon and can look down that road, field or peak into that group of buildings a kilometer away. The lightweight armor of *Air*-Mech organizations absolutely requires an expanded over-match in tactical awareness gained down to the platoon in order to effectively mitigate the absence of heavy armored plate. Downlinks from higher headquarters will not be enough. Company commanders and Platoon Leaders and their Sergeants will need their own mini-systems to "*bird dog*" the tree lines, reverse slopes and building clusters in order to survive against enemy heavy armor. The lethality of lightweight, "point and shoot" anti-tank systems like Javelin and LOSAT need a minute or two ambush set-up time in order to have a terrain defilade position to protect themselves from enemy counter-fires. Even the Mobile Gun System or Armored Gun System vehicles in our proposed AMS force structure, while able to point/shoot/kill enemy armor with a large caliber, high-pressure gun to prevail in a surprise meeting engagement (Lesson learned from the 9th Division High Technology Test Bed experiments in the 1980s) are best employed in ambush positions to protect them from enemy fires.

4. Destroy Majority of Massed Enemy Armor with Indirect-Precision Fires: Using the information gained through battlefield digitalization, the *Air*-Mech-Strike Commander must strive to defeat 80 percent or better of his opposing force through over-the-horizon indirect fires from precision munitions like top-attack, Enhanced Fiber Optic Guided Missiles, guided 120mm mortars and air/ground launched millimeter-wave *Hellfire* ATGMs. The small-arms and RPG protective level armor on *Air*-Mech vehicles necessary to achieve helo-transportability makes it important to have good tactical planning based on timely reconnaissance to avoid surprise, massed vehicle direct-fire fights. Direct firefights will not be eliminated by *Air*-Mech-Strike but their occurrence should be an anticipated event where the conditions for a quick-kill using all of the weaponry of the AMS force, Army and joint fires have been set.

Today's Divisions are designed for the opposite results. The majority of an U.S. Army's heavy Division firepower is housed in the direct-fire M1 and M2 tank and infantry fighting vehicles. Even the Divisional artillery has relatively short-range howitzers intended to directly support the maneuver of these direct-fire systems rather than reaching out to destroy the enemy in depth. The current attack aviation assets in the Divisional Aviation Brigade are likewise intended to set the conditions for the closure with the enemy of the M1/M2 equipped Brigades. This singular course of action changes under *Air*-Mech-Strike. Artillery and platforms of indirect fire precision munitions can become the main effort in destroying massed enemy formations if the situation dictates. Direct fire in these situations would be used to finish-off precision strike survivors and complete the destruction of the enemy as a fighting organization.

Its important to note that this Axiom is addressing massed armored formations and groupings. Close-in fighting in complex and urban terrain will still have a high degree of direct-fire mainly from gun-equipped *Air*-Mech fighting vehicles and their squads of infantry clearing buildings, trench lines or rooting out a dispersed enemy tank platoon in a wooded area. In these situations, the effectiveness of long-range precision munitions is diminished and unlikely to be the decisive defeat mechanism, though they can be used at specific rooftop targets. The *Air*-Mech-Strike force has a high percentage of infantry forces and with its small lightweight vehicles will be more effective at maneuvering in the complex and urban terrain to effect close-range direct-fire fights than units without any armored vehicles by force structure or circumstance if they are found to be too big to maneuver in the space available. Direct fire is not going away under *Air*-Mech-Strike but the sole dependence on direct line-of-sight engagements as the only kill mechanism will be expanded to include indirect, non-line-of-sight fires.

5. All Mode Refuel-able Aircraft: The *Air*-Mech-Strike concept places an extreme premium on the range of Army Helicopters. Current mechanized maneuver places Army helicopters in close support roles with only a few Corps-level missions directed against deep targets. Manufacturers were directed to emphasize lift capacity over speed and range. The greater the striking range of an *Air*-Mech-Strike, the greater the chances for achieving surprise and ultimately mission success through rapid positional advantage. The ability to refuel quickly and under combat conditions is therefore a specified Axiom under *Air*-Mech-Strike. Additional fuel tanks can and should be retrofitted to the current UH-60 and CH-47 fleets however; new and creative ways to effect forward refueling will still be required.

Army helicopters must be able to refuel on board naval vessels. Currently the tailwheel of a UH-60 is a few feet too aft of the tail to allow landing and hence refueling off of most small Navy vessels. Army helicopters must be fully able to land and refuel off of any Navy ship. *Air*-Mech-Strike capable *Blackhawks* and *Chinooks* will have crews trained and qualified in shipboard operations and equipped with rotor brakes, landing gear and any other equipment required. Additional refueling methods include proficiency training in USAF C-130/17 wet-wing refueling operations and the most challenging of all, mid-air refueling. Other creative methods to extend the range of an *Air*-Mech-Strike is the ability to draw fuel from un-manned refuel blivets parachuted into forward locations by USAF Aircraft. The latter system would have a tall refuel pipe where an Army helicopter could hover over to and draw fuel from a mid-air refueling nozzle without ever actually landing and getting out of the aircraft termed "*Humming-Bird Refueling*" (HBR). In wartime, anti-tampering devices would be fitted to the blivets that would prevent an enemy from using these blivets.

CH-47s with internal fuels tanks performing "*Fat Cow*" forward refueling would be common place in an *Air*-Mech-Strike operation. Forward refueling and the relatively short-range legs of the current aircraft fleet are a problem area of the *Air*-Mech-Strike concept but can be mitigated by an aggressive all-mode refueling training and equipping program.

6. 24-Hour Aviation Operations: Aircraft do not get tired but aircrews do. At least two crews are needed to double the daily sortie rate of aircraft fleets to better support a typical *Air*-Mech-Strike operation. It's a lot easier to transport crews than aircraft in a forward theater of operation. This would be an excellent multi-component training opportunity where the extra pilots could be monthly reservists assigned to an active component unit. History has demonstrated that reserve pilots are as proficient or in many cases more experienced than their active counter-parts. Reserve units however, often lack the time to progress toward large-scale collective training.

Not only will an aviation unit's sortie rate need to increase to effect the actual *Air*-Mech-Strike, but the follow-on logistical support to the inserted forces will have to be maintained largely by aerial resupply from these same helicopters. The current "*cold war*" crewing of one crew per aircraft mean than an airframe sits on the ground for half a day while the single crew rests.

7. Theater Support Base Logistics: Air-Mech-Strike requires an assault or expeditionary mind set where the traditional "*bring-everything*" way of supporting an Army deployment is replaced by a surge "*live-out-of-ruck*" for 45-days mentality. The Theater Support Base (TSB) normally planned in a relatively secure allied nation like Italy for the Yugoslavia campaign or Japan for Korea would be the primary logistic activity area. *Air*-Mech-Strike operations are more effective if launched from TSBs which can only be gained if the "*legs*" range of Army helicopters can be increased. The TSB would be the main replacement transfer point for major end items like vehicles and aircraft that can not be fixed by simple replacement or "*tool box*" maintenance procedures. The logistic foot-print in the maneuver box will be only that which is critical to feeding, refueling and rearming. Most maintenance and other support activities will be done at the TSB. The TSB would push critical daily supplies and whole end items as needed through daily flights of C-130s and C-17s to the ISB and or FSBs. Daily cargo aircraft flights to the United States mainland would be used to exchange aircraft and other major end items. The end state logistic program is to support from a TSB via air lines of communication a fast-moving force down range that is surging to achieve a decisive action within 45 days. By which time a stabilizing heavier 2-D force can them reinforce the *Air*-Mech 3-D Force if required.

8. Split-Loading: *Air*-Mech-Strike requires vehicles and other follow-on equipment to be capable of splitting their loads in order to distribute the payload over more helo-sorties as a remedy for dealing with higher elevations, higher temperatures and or longer ranges. Separable modular armored tubs are a strong concept contender to achieve this goal. Split-loading additional add-on armor, weapons, and other mission equipment carried by other aircraft could also be done to enhance capabilities and lessen vulnerabilities. *Air*-Mech force vehicles will be fairly small and offer only sufficient room for an assault-equipped infantry Soldier. Therefore, two and four wheel support trailers/carts will need to be utilized to carry rucksacks and additional gear. The support trailers/carts would be initially left behind in the assault phase then flown in as split loads to their vehicles after. The four wheel feature allows multiple trailers/carts to be linked together and moved by a single prime mover to expedite transport forward.

9. Air Parity as Minimum: *Air*-Mech-Strike needs at least air parity within the corridors its conducting air maneuver. It does not require air supremacy or overall superiority. History has shown that helicopters and low-flying fixed-wing aircraft can operate effectively under the sorties of enemy fighter aircraft. Army helicopters and USAF A-10s are already equipped with air-to-air missiles and the introduction of the AH-64D *Longbow Apache* with its counter-air radar mode will make the engagement of U.S. Army helicopters from enemy fighters all the more difficult. The AH-64 will in effect become a flying "all weather SAM battery". During *Desert Storm*, Iraqi *Hind* and *Hip* Helicopters flew many sorties up the front in daylight with few losses under coalition air supremacy. A-10s capable of escorting USAF transports and Army helicopters succeeded in shooting down enemy aircraft that got in their way during the Gulf War.

If the enemy has gained local air superiority or worse, then continued *Air*-Mech-Strike tactical movements are not likely to succeed primarily from the vulnerability of helicopter pick-up zone marshaling, landing and refueling areas. What sets the AMS force apart from other Air Assault force structures which would become defensive/stagnant when they lose their air 3-D maneuver component to weather or enemy action etc., is that the AMS force can maneuver in a dispersed, mechanized 2-D fashion and continue the mission. The aircraft will laager farther back at the TSB and await the regaining of at least air parity.

10. Two-to-One Vehicle-to-Aircraft Ratio: There are two conflicting design mission criteria that challenges the *Air*-Mech-Strike concept. On the one hand, the lifted vehicle should be as heavy as possible using this weight for survival enhancing armor and greater lethality weapon systems and not squandering it away in ground propulsion means. On the other hand, the lifted vehicle should be as light as possible to support multiple systems being lifted on one aircraft to achieve a force of sufficient size to dominate the battle and accomplish the mission. Its likely that aircraft designers will push for a minimal "*least-vehicle*" solution and the armor manufacturers will opt for a robust airframe to haul their "*maximum vehicle*" solution.

The expense of a vertical take-off and landing aircraft is the critical mission design criteria. The optimum axiom appears to be the two-vehicles-per-aircraft lift ratio with a small number of one-to-one situations for a limited number of high priority vehicles. This means that a "*killer*" vehicle designed to shoot-on-the-move and employ a larger weapon system and some additional armor would be the high-priority single-lifted vehicles and a lightweight personnel carrier variant weighing half as much would be the majority. This axiom is already partially violated with the UH-60 which can only carry one *Wiesel*-based vehicle. However, the CH-47 more accurately portrays the axiom by being able to lift one *Gavin* Fighting Vehicle (M113A3 based) or two *Wiesel* RFVs.

The USAF C-130H/J fixed-wing STOL aircraft achieves the 2-for-1 axiom in that it can carry either 2 *Air*-Mech "APC" vehicles (GFV and a RFV or 2 RFVs) or a single "killer" M8 Armored Gun System. These can be parachute air-dropped or short-field airlanded.

11. Streamlined External Loads (SEL): In order to achieve an increase in the range of Army helicopters to support deep *Air*-Mech-Strikes, a rolling take-off from a semi-prepared field would be extremely advantageous. This would allow greater aircraft gross weights and larger payloads. To achieve this near-term the UH-60L and CH-47F could be modified with kits to extend their landing gears to enable Streamlined External Loads (SEL) like the current UH-60L *Firehawk* firefighting fuselage kit. The *Firehawk* has an externally streamlined load of a large water tub that fits close to the underside of the fuselage and uses extended gear to achieve rolling take-offs if necessary. The SEL concept also allows much closer terrain flight than current sling-loading which enhances survivability and allows for greater forward speeds by eliminating the pendulum effect which in turn increases range.

12. Dedicated DEAD-LE Force Commander: *Air*-Mech-Strike operations will require the formation of a flying Destruction of Enemy Air Defense Lanes Enroute (DEAD-LE) Commander. The DEAD-LE Commander would command a force of helicopters and a dedicated staff with only one purpose to identify and destroy enemy air defenses. UH-60 aircraft would be utilized to form an Army version of the USAF "*Wild Weasel*"

Airland: Mobile Gun System rolling off C-130 *Hercules* (U.S. Army)

units with radar detection equipment, jammers, mass chaff dispensers, IR smoke screen laying devices, anti-radar missiles, EOCM laser weaponry, ADA killing UAVs and close coordination with joint counter ADA agencies. The demands of *Air*-Mech-Strike and the increase in the lethality of air defense systems world-wide have rendered the current Suppression of Enemy Air Defense (SEAD) planning done by the unit fire support officer inadequate. The goal must be destruction—not just suppression and that requires a dedicated staff with both intelligence and killing hardware systems to detect and destroy enemy air defense systems.

Chapter 6

Air-Mech-Strike is Possible Today

"The aim of bringing together the story of Air Mobility is to afford a better understanding of where we are at present and to make sure that the surprisingly successful use of the third dimension to fight the ground war in the past remains a matter of record when we consider where we go from here."
—General John R. Galvin (USA, Retired), page XI;
***Air Assault**: The development of Air-mobile Warfare*

Kosovo, March 1999: Helicopters can Successfully Penetrate Modern Air Defenses

The U.S. Air Force 16th Special Operations Wing after only three days of being forward-deployed, took to the skies as part of a joint team tasked to rescue a downed F-117 *Nighthawk* stealth fighter pilot. After learning of the downed F-117 pilot, a special operations formation of MH-53J *Pave Low III*s and MH-60G *Pave Hawk* helicopters was launched. After the flight conducted a low-level, Night-Vision Goggle (NVG) air refueling, without communication or lighting with an MC-130P *Hercules* tanker, the confirmed location was relayed to the rescue force. The downed American pilot was within close proximity of downtown Belgrade, the Serb capitol city—a Serbian MIG fighter base, and within a few miles of three enemy ground force Brigades. The rescue force discovered the downed pilot near a major road where vehicles were stopping to dismount Soldiers and search dogs. An MH-60G *Pave Hawk* assigned to the 55th Special Operations Squadron landed within 100 yards of the survivor. The pilot was on board within a few seconds and the flight departed the area. The flight returned safely five and a half hours after initial takeoff.

On the second rescue when an F-16 was downed, another special operations helicopter rescue force was assembled and

117

launched to recover the pilot. During this operation the moon was nearly full, giving an advantage to enemy air defenses. The formation dodged one SA-9 and two SA-6 surface-to-air missiles. The MH-60G *Pave Hawks* dispatched to pick up the downed pilot sustained significant damage from small-arms fire. Though the route home was strewn with Serbian forces using anti-aircraft fire, searchlights and small-arms, they returned safely just as daybreak began to light the formation.

U.S. Special Operations Command Commanding General Peter Schoomaker stated at a ceremony honoring America's helicopter heroes:

"The real success story here is not in the high-speed bells, whistles, gadgets and devices these crews used, but rather the living manifestation of the legacy of the warrior spirit".[1]

U.S. has the Best Pilots and Aircraft in the World

It is clear that America's Army and Air Force have the best aircraft pilots in the world and that they can penetrate enemy air defenses successfully to insert/extract personnel or conduct attack missions. In 1983 during the invasion of Grenada, USAF MC-130 *Combat Talon* (Special Operations version of the C-130 *Hercules*) pilots on-the-scene decided to drop down to under 500 feet to parachute Army Rangers under the depression ability of Cuban anti-aircraft guns, winning the fight to take Point Salines airfield. In Panama, USAF aircrews flew through enemy tracer fires criss-crossing the night to airdrop U.S. Army Rangers and 82d Airborne Paratroopers, to include M151 "*Gun jeeps*" and M551 *Sheridan* light tanks. A critical event in the 1990 air campaign in *Desert Storm* began when U.S. Army *Apache* attack helicopter crews destroyed key Iraqi ground radar sites to "open the door" for the waves of fixed-wing aircraft to follow. Army aircrews are required to be Night Vision Goggle qualified to fly Nap-Of-the-Earth (NOE) using the terrain and vegetation to mask their aircraft from enemy detection and fire. Terrain contour, night-adverse weather flight, and chaff, jammer penetration aids are all employed to penetrate even the most heavily defended air spaces.[2] There are technological advances like EOCM laser weaponry that can be aircraft mounted to counter the threat of MAN-Portable Air Defense Systems (MANPADs). In the RAND Report, *Enhancing Air Power's Contribution Against Light Infantry Targets*, Alan Vick, David T. Orletsky, John Bordeaux, and David A. Shlapak conclude:

> *Light infantry forces produce signatures that can be detected by airborne and air-implanted ground sensors. A multiphenomenology approach—combining many different sensor types—offers the highest probability of detecting and identifying enemy light forces. Electro-optical sensors on platforms flying at 5,000 feet (ft) above ground level (AGL) and within 2 to 3 kilometers (km) of the target are key to distinguishing between enemy forces and noncombatants and/or friendly forces.*[3]

Its apparent that U.S. fixed-wing and helicopter pilots have the necessary courage for "cross-flot" operations. With the largest fixed-wing (USAF) and helicopter fleet in the world, coupled with new technologies, the U.S. Army is able to fully exploit this 3-D maneuver.

U.S. has the Most Accurate Army and Joint Fires in the World to Support *Air*-Mech 3-D Maneuver

The Gulf War proved that the U.S. military has the world's most precise array of precision weaponry and can mass these fires for decisive effects. This reality dates back to World War II, where the Germans complained that the American artillery, once it got a target, killed it with overwhelming speed. Our Asian enemies in later conflicts died by the thousands from this firepower. History has shown that despite all these tremendous effects of fire, we cannot expect these fires to defeat the enemy single-handedly. In the recent air campaign in Kosovo, after weeks of around-the-clock bombing, few of the Serbian Army ground targets were destroyed. Their capitulation to NATO demands came only after they saw NATO was preparing for an Army ground offensive.[4] Inserting *Air*-Mech-Strike forces on the ground to gain positional advantage, to observe and employ organic weaponry or call-in fires can unhinge the enemy for our 2-D ground forces to maneuver. With *Air*-Mech armored vehicles, U.S. Army forces will have an additional advantage of protection and firepower.

Using the Proper Mode of Maneuver: Helicopter Difficulties Avoided

There are essentially 3 basic forms of Army maneuver; mounted in aircraft, mounted in ground vehicles or dismounted on foot. Commanders must strive to use the best possible maneuver form for a given tactical situation. Their options available must include as many tools as possible; be it parachuting, airlanding, hovering and fast roping or rappelling from aircraft, or using ground vehicles. In Vietnam, and continuing to the present day, U.S. light forces only have the option of going from Air-mounted to dismounted (or vise-a-versa)—since aircraft have to return to a safe area to refuel and re-arm—this makes planners have to deliver infantry close to or on the objective where the enemy is often able to fight back and the result has been helicopter losses in battles like *Koh Tang* island in 1975.[5] By not fielding Airborne and Air Assault forces equipped with all 3 maneuver forms that could be used inter-changeably as the situation dictates, we have at times seen maneuver mis-matches that have put our Soldiers and the mission at risk.[6] *Air*-Mech-Strike solves this once and for all by giving our 3-D forces the means to fully maneuver at all times and in all terrain.

Helicopter Slingload and Internal Load Procedures already in Place to Transport *Gavins/Ridgways*

The U.S. Army already owns 6 *Wiesel 1* AFVs that were tested extensively by the Army's TEXCOM agency to include fixed-wing aircraft parachute airdrop and helicopter sling-load procedures[7]. Some are used at the McKenna MOUT facility at Fort Benning, Georgia where their small size allows operations throughout the village. The Germans have been using *Wiesels* for almost a decade in U.S. made CH-53G helicopters and have airdropped them from as low as 400 feet successfully.[8] The precise airdrop/slingload techniques for American *Ridgway* Fighting Vehicles based on the German *Wiesel* AFV would only require a minor series of certification tests as the lion's share of the work is already done. German Army Airborne *Wiesels* on exchange duty are used routinely at the Joint Readiness Training Center at Fort Polk, Louisiana[9]. The reports from dozens of U.S. Soldiers is they eagerly await their fielding in the U.S. Army[10] Wiesels in U.S. Army service with rubber band tracks could easily drive on roads without need of truck transporters achieving wheeled-vehicle like speeds[11].

M113 type AFVs have been CH-47 helicopter sling-loaded in combat conditions in Vietnam. Ralph Zumbro writes in *Lighten Up, Guys*, page 25, of U.S. Army Armor Magazine, November-December 1999 issue;

"*Back during the Vietnam War, the 25th Division found out that a CH-47 can lift an ACAV and move it across about 20 miles of battlefield.*"

More recently, Singapore Army improved model CH-47Ds were photographed lifting M113s for short distances.

Para-Mech Airborne Operations

The U.S. Army has the world's most skilled parachute Riggers—this large body of experts has for years routinely "heavy-dropped" the M551 *Sheridan* light tank at 42,000 pounds from C-130s and constantly improves its techniques and equipment.[12] Loads now upon landing have their parachutes release to prevent tip-overs and wind dragging. Guided parafoil systems have been fielded for pinpoint delivery of loads up to armored vehicle sizes from aircraft standoff ranges.[13] These precision parachute systems offer exciting near-term possibilities for para-mech missions. Rigging procedures are already in place to parachute Low-Velocity AirDrop (LVAD) *Gavins/ Ridgways*.[14]

The official manual for LVAD rigging the GFV is Field Manual 10-567, TECHNICAL ORDER NO. 13C7-16-171 HEADQUARTERS DEPARTMENT OF THE ARMY DEPARTMENT OF THE AIR FORCE Washington, DC, 29 June 1979 and 4 August 1997

AIRDROP OF SUPPLIES AND EQUIPMENT: RIGGING TRACKED PERSONNEL-CARGO CARRIERS describes in detail how to set-up M113s for parachute air-delivery. Closer examination of this manual shows its actually takes less steps and man-hours to rig a 22,000 pound M113 for airdrop than it is to dismantle and rig for airdrop a 22,000 pound 5-ton capacity truck commonly employed by Army Airborne units.[15]

The Germans have been using *Wiesels* for almost a decade and have airdropped them from as low as 400 feet successfully. The Airborne/Special Operations Test Board at Fort Bragg, North Carolina has certified them for airdrop using U.S. airdrop parachutes and materials.

Tactics, Techniques and Procedures already in Place for *Gavins*/Mobile Gun Systems

There is still much institutional knowledge on deploying a Mechanized Infantry Battalion in M113A3s; there are still National Guard units and Active army units on peacekeeping duty in Macedonia in the Balkans using the *Gavin*. FM 7-7 *Mechanized Infantry Platoon and Squad* is widely available in paper form and viewable online at the Army Digital Training Library.[16]

Likewise, the U.S. Army wrote FM 17-18 *Light Armor Operations* before the M8 Armored Gun System was cancelled. This manual, like FM 7-7 is still available in paper and online version.[17]

A unit trained on the complex BFV will find transitioning to the simpler *Gavin* easy to master; National Guard light infantry units with no mechanized experience whatsoever are routinely given "OJT" and placed in loaner M113A2s visually modified to look like BMPs to fight as OPFOR augmentees at the National Training Center. These OPFOR mech-infantry Soldiers do very well against their more sophisticated equipped BlueFor foes at the NTC. At the JRTC at Fort Polk, Louisiana, active-U.S. Army Airborne Combat Engineers were recently outfitted in modern M113A3s on loan from the attached National Guard unit and did very well with them; as they provided armored mobility, increased .50 caliber Heavy Machine Gun firepower and a provided a means to tow heavy equipment like MICLIC rocket mine clearing trailers.[18]

Its clear that the Soldiers, equipment and doctrine are already in place to effect *Air*-Mech-Strike warfare possible today.

Air-Mech-Strike Reconnaissance

Napoleon once said: *"Time spent in reconnaissance is never wasted".* The Center for Army Lessons Learned (CALL), National Training Center (NTC), Joint Readiness Training Center (JRTC) and the Combat Maneuver Training Center (CMTC) have directly linked reconnaissance failures to overall mission failure. Successful reconnaissance efforts often result in overall battle success. Seeing the enemy and the terrain is essential to a Comander's decision cycle on the battlefield. Real time intelligence is what's critical behind any reconnaissance effort. Technology enhancements such as UAVs and satellites, have provided unmanned means to gather pictures of the terrain and enemy, however, a picture doesn't always provide the adversary's intentions on contact, his level of training, morale, and trafficability of the terrain. These factors are best derived from on-the-ground information; in the form of manned reconnaissance vehicles, dismounted scouts and NOE helicopters. The combination of high-tech sensors and human intelligence, working in a continuous reconnaissance effort, is the ideal situation. The *Air*-Mech-Strike concept greatly enhances the human component of this reconnaissance effort by leveraging the speed of Air Assault maneuver with detailed terrain knowledge gained by ground light-armored vehicle recon teams. A Reconnaissance Surveillance Targeting and Acquisition (RSTA) organization based on the *Air*-Mech-Strike model, with its large number of helo-transportable armored platforms can achieve far greater depth, terrain coverage and speed of recon maneuver over our current recon deficient organizations.

Reconnaissance Force Needs Fast Entry into Theater of Operation

The AMS RSTA model can be delivered into any theater support base using five Boeing 747 sorties for the entire ground component and about five to six C-17s for a UH-60 helicopter task force to provide platoon size lifts. If funding were available for pre-lease agreements, it would be possible to replace the five C-17s with open cargo-nose Boeing 747s making the entire AMS RSTA completely transportable by commercial aircraft. The need for extensive material handling equipment on the receiving end at the theater support base would be a factor to consider in the use of use of commercial aircraft. Current force models and those based on 20,000 pound armored cars do not have this option and must be deployed by C-17/C-5. The AMS RSTA model also has the capability to quickly enter a theater of operation from a nearby supporting country. An AMS RSTA could be lifted by UH-60 and CH-47s from Italy into many parts of Yugoslavia, or from Japan into Korea as examples. The lightweight nature of the AMS vehicles also facilitates large volume "para-mech" from C-5/C-17/C-141 and C-130 aircraft. Such rapid insertion speeds for a robust RSTA force enhance the success of the deployment of the main body into theater.

Reconnaissance; a Continuous Priority

Reconnaissance operations are normally characterized by continuous 24 hour-a-day activity over great spans of territory. This places a taxing demand on the logistic support structure and the personnel work load. The AMS model mitigates these challenges by having vehicles that use very little fuel, have low maintenance and need only the support of another like vehicle or at worse a HMMWV wrecker. The large number of platforms made possible by using light weight vehicles also means that there are more recon assets to distribute the required tasks thus lessening the direct Soldier work load.

Reconnaissance Demands Wide Area Coverage

Proper reconnaissance and surveillance provides information throughout the battlespace to the Commander. America's potential adversaries are well aware of our capability to direct precision firepower on what we can see. To survive, they will endeavor to occupy and operate out of difficult terrain in forested, mountainous and urbanized areas. The reconnaissance task therefore requires speedily access to and mobility in restricted terrain. The enemy must be denied any sanctuary. AMS offers the best solution to this challenge by employing large numbers of small agile mechanized platforms in the form of the *Wiesel*-based *Ridgway* Fighting Vehicle and the commercial-based ATV. Access to these difficult areas is not only a function of employing smaller or agile vehicles but also in the AMS capability to vertically insert these platforms into remote areas. Currently most conventional reconnaissance efforts in difficult terrain are addressed with dismounted recon teams that employ hand held observation equipment without weapons other than those for personal protection. The AMS RSTA model corrects this deficiency by providing the employment of more capable recon and surveillance equipment and significant increase in weapon lethality. The result is that an AMS RSTA organization can cover a far greater area with the large number of light weight platforms and with terrain access that our current capability cannot provide.

Speed of Reconnaissance

Sudden changes in the operational situation often occur, such as the recent sprint of Russian Paratroopers by land from Bosnia to seize the Pristina Airport in Kosovo in 1999. The AMS RSTA has the capability to rapidly displace via helicopters over natural and man-made obstacles to react to these changes. The Commander can quickly gain situational awareness on a new development with his 3-Dimensional AMS RSTA organization.

Air-Mech-Strike is the most Cost-Effective Way to Transform the U.S. Army for Full-Spectrum Operations

The U.S. Army's heavy units are already 50% composed of M113A3-type tracked AFVs which have modification potential for firepower, add-on armor and require no truck transporters. The Army has thousands of older model M113s that can be converted to A3 standard at low cost. With band-tracks and other lightweight accessories the M113A3 *Gavin* can fly by CH-

47F helicopters. The M113A3 lightened under 10-tons can be *Air*-Meched two at a time in a C-130.

COST COMPARISON M113A3 BRIGDE VESUS FROM-SCRATCH ARMORED CAR BRIGADE

Build-on the M113A3 Brigade Combat Team model

180 M113A2s converted to M113A3 standards @ $150,000 each = $27,000,000 MILLION
60 *Ridgways* based on *Wiesel 2s* @ $400,000 each = 24,000,000 MILLION
TOTAL $51.0 MILLION

Build-from-scratch Brigade with New Purchase Armored Vehicles

400 LAV-type armored cars @ $880,000 each = $ 352,000,000 MILLION
TOTAL $352.0 MILLION

    Hidden costs are the time loss and hiring of civilian contractors to train thousands of Army Soldiers on the new armored cars/trucks. In contrast, the M113A3 is already in Army service, and Soldiers are trained to operate and maintain them, Class IX repair parts are already produced around the world. The "TMs" are already widely distributed throughout the U.S. Army.
    Instead of an entire U.S. Army Brigade taken out of the deployment cycle putting an even greater burden on those remaining on mission-available status to meet the ever increasing OPTEMPO, with *Air*-Mech-Strike, only 1 maneuver Battalion of the Brigade Combat Team is effected with a transformation to the capable M113A3 *Gavin* and the creation of the RSTA *Ridgway* Squadron.

M113A3 *Gavin/Ridgway* Brigade Transformation Time

50% of a Heavy Brigade is already in M113A3s, no change required

2 Battalions of the remaining 50% of the Brigade receives New Equipment Training (NET) of the already-in-service Army M113A3 vehicle and the new *Ridgway* = time loss no more than 30 days

250 Drivers to be trained @ 8 hours a day = 1600 man-hours of training

1600 hours for a 5-day work week = 8000 man-hours

4 weeks x 8000 man-hours = 32,000 man-hours to train Drivers

250 *Gavin/Ridgway* Commanders trained with their Drivers @ 8 hours a day = 1600 man-hours of training

1600 hours for a 5-day work week = 8000 man-hours

4 weeks x 8000 man-hours = 32,000 man-hours to train *Gavin/Ridgway* Commanders

TOTAL = 64,000 man-hours to train 250 *Gavin/Ridgway* Crews to make a Brigade Combat Team

Armored Car Brigade Conversion Time

100% of the BCT requires armored car New Equipment Training (NET)

400 Drivers to be trained @ 8 hours a day = 3200 man-hours of training

3200 hours for a 5-day work week = 16, 000 man-hours

4 weeks x 16, 000 man-hours = 64,000 man-hours to train Drivers

200 Wheeled Armored Car Commanders trained with their Drivers @ 8 hours a day = 3200 man-3200 hours for a 5-day work week = 16, 000 man-hours

4 weeks x 16, 000 man-hours = 64, 000 man-hours to train Wheeled Armored Car Commanders

TOTAL = 128,000 man-hours to train 400 Wheeled Armored Car Crews to make a Brigade Combat Team

Summary

The technology to effect *Air*-Mech-Strike 3-D maneuver warfare is of sufficient maturity to field the capability today. We can buy German *Wiesels* (*Ridgway* Fighting Vehicles) "*off-the-*

shelf since the German Army has already perfected the design which is in production. The M113A3 *Gavin* AFV is a proven vehicle in the U.S. Army. The U.S. Army already has the world's largest and most capable helicopter force. Most of the Equipment, TTP, institutional knowledge, trained combat leaders, spare parts are in hand to effect *Air*-Mech-Strike TODAY. With modification for *Air*-Mech-Strike operations, all of the 3-D parts are C-130 or CH-47 transportable. The smaller 3-D parts are UH-60L *Air*-Mech-able. All of the 2-D parts are C-17 *Air*-Mech-able. The Army can combine these elements into a stronger whole for an INTEGRATED *Air*-Mech-Strike capability. Given the state of the current helicopter fleets and light *Air*-Mech vehicles, the *Air*-Mech-Strike concept is at a stage of development similar to where mechanized warfare was in the early 1940s. There are many technical and operational employment challenges ahead. We should not repeat the mistakes of the inter-war years with the U.S. Army that missed great opportunities to develop the tank and the associated infantry armored personnel carrier. To commit resources to field another 2-D type—force will only forestall the development of 3-D maneuver warfare.[19]

---

[1] U.S. Air Force 16th Special Operations Wing, (Hurlburt Field, Florida: *Commando Online* newsletter)
http://www.hurlburt.af.mil/commando/archives/
[2] U.S. Army Military Academy Aviation Home page: Nap-of-the-Earth *http://www.dmi.usma.edu/Branch/AV/Aviation.htm*
[3] *http://www.rand.org/publications/MR/MR697/*
[4] U.S. Air Force Association Air Force Magazine online:
http://www.afa.org/magazine/world/1199world.html
[General] *Clark also asserted, without offering evidence, that another factor influencing Milosevic's decision to capitulate was that "he had ample evidence to conclude that, had he not conceded when he did, the next step would have been the long-awaited and much-talked-about NATO ground effort."*
[5] Richard Pyle, *U.S. Team Seeking Remains of 18 Servicemen Killed in 1975* (New York.: Associated Press, November. 2, 1995)
Koh Tang debacle: At 11:21 a.m. on May 12, the U.S. mer-

chant ship MAYAGUEZ was seized by the Khmer Rouge in the Gulf of Siam about 60 miles from the Cambodian coastline and eight miles from Poulo Wai island. The ship, owned by Sea-Land Corporation, was en route to Sattahip, Thailand from Hong Kong, carrying a non-arms cargo for military bases in Thailand. Capt. Charles T. Miller, a veteran of more than 40 years at sea, was on the bridge. He had steered the ship within the boundaries of international waters, but the Cambodians had recently claimed territorial waters 90 miles from the coast of Cambodia. The thirty-nine seamen aboard were taken prisoner.

President Ford ordered the aircraft carrier USS CORAL SEA, the guided missile destroyer USS HENRY B. WILSON and the USS HOLT to the area of seizure. By night, a U.S. reconnaissance aircraft located the MAYAGUEZ at anchor off Poulo Wai island. Plans were made to rescue the crew. A battalion landing team of 1,100 marines was ordered flown from bases in Okinawa and the Philippines to assemble at Utapao, Thailand in preparation for the assault.

The first casualties of the effort to free the MAYAGUEZ are recorded on May 13 when a helicopter carrying Air Force security team personnel crashed en route to Utapao, killing all 23 aboard.

Early in the morning of May 13, the Mayaguez was ordered to head for Koh Tang island. Its crew was loaded aboard a Thai fishing boat and taken first to Koh Tang, then to the mainland city of Kompong Song, then to Rong San Lem island. U.S. intelligence had observed a cove with considerable activity on the island of Koh Tang, a small five-mile long island about 35 miles off the coast of Cambodia southwest of the city of Sihanoukville (Kampong Saom), and believed that some of the crew might be held there. They also knew of the Thai fishing boat, and had observed what appeared to be caucasians aboard it, but it could not be determined if some or all of the crew was aboard.

The USS HOLT was ordered to seize and secure the MAYAGUEZ, still anchored off Koh Tang. Marines were to land on the island and rescue any of the crew. Navy jets from the USS CORAL SEA were to make four strikes on military installments on the Cambodian mainland.

On May 15, the first wave of 179 marines headed for the island aboard eight Air Force "Jolly Green Giant" helicopters. Three Air Force helicopters unloaded marines from the 1st Battalion, 4th marines onto the landing pad of the USS HOLT and then headed back to Utapao to pick up the second wave of marines. Planes dropped tear gas on the MAYAGUEZ, and the USS HOLT pulled up along side the vessel and the marines stormed aboard. The MAYAGUEZ was deserted.

Simultaneously, the marines of the 2/9 were making their landings on two other areas of the island. The eastern landing zone was on the cove side where the Cambodian compound was located. The western landing zone was a narrow spit of beach about 500 feet behind the compound on the other side of the island. The marines hoped to surround the compound.

As the first troops began to unload on both beaches, the Cambodians opened fire. On the western beach, one helicopter was hit and flew off crippled, to ditch in the ocean about 1 mile away. The pilot had just disembarked his passengers, and he was rescued at sea.

Meanwhile, the eastern landing zone had become a disaster. The first two helicopters landing were met by enemy fire. Ground commander, (now) Col. Randall W. Austin had been told to expect between 20 and 40 Khmer Rouge Soldiers on the island. Instead, between 150 and 200 were encountered. First, Lt. John Shramm's helicopter tore apart and crashed into the surf after the rotor system was hit. All aboard made a dash for the tree line on the beach.

One CH-53A helicopter was flown by U.S. Air Force Major Howard Corson and 2Lt. Richard Van de Geer and carrying 23 marines and 2 U.S. Navy corpsmen, all from the 2nd Battalion, 9th marines. As the helicopter approached the island, it was caught in a cross fire and hit by a rocket. The severely damaged helicopter crashed into the sea just off the coast of the island and exploded. To avoid enemy fire, survivors were forced to swim out to sea for rescue. Twelve aboard, including Maj. Corson, were rescued. Those missing from the helicopter were 2Lt. Richard Van de Geer, PFC Daniel A. Benedett, PFC Lynn Blessing, PFC Walter Boyd, Lcpl. Gregory S. Copenhaver, Lcpl. Andres Garcia, PFC James J. Jacques, PFC James R. Maxwell, PFC Richard W. Rivenburgh, PFC Antonio R. Sandoval, PFC Kelton R. Turner, all U.S. marines. Also missing were HM1 Bernard Gause, Jr. and HM Ronald J. Manning, the two corpsmen.

Other helicopters were more successful in landing their passengers. One CH-53A, however was not. SSgt. Elwood E. Rumbaugh's aircraft was near the coastline when it was shot down. Rumbaugh is the only missing man from the aircraft. The passengers were safely extracted. (It is not known whether the passengers went down with the aircraft or whether they were rescued from the island.)

By midmorning, when the Cambodians on the mainland began receiving reports of the assault, they ordered the crew of the MAYAGUEZ on a Thai boat, and then left. The MAYAGUEZ crew was recovered by the USS WILSON before the second wave of Marines was deployed, but **the second wave was ordered to attack anyway**.

Late in the afternoon, the assault force had consolidated its position on the western landing zone and the eastern landing zone was evacuated at 6:00 p.m. By the end of the 14-hour operation, most of the marines were extracted from the island safely, with 50 wounded. Lcpl. Ashton Loney had been killed by enemy fire, but **his body could not be recovered**.

Protecting the perimeter during the final evacuation was the machine gun squad of PFC Gary L. Hall, Lcpl. Joseph N. Hargrove and Pvt. Danny G. Marshall. They had run out of ammunition and were ordered to evacuate on the last helicopter. It was their last contact. Maj. McNemar and Maj. James H. Davis made a final sweep of the beach before boarding the helicopter and **were unable to locate them**. They were declared Missing in Action.

The eighteen men missing from the MAYAGUEZ incident are listed among the missing from the Vietnam war. Although authorities believe that there are perhaps hundreds of American prisoners still alive in Southeast Asia from the war, most are pessimistic about the fates of those captured by the Khmer Rouge.

In 1988, the communist government of Kampuchea (Cambodia) announced that it wished to return the remains of several dozen Americans to the United States. (In fact, the number was higher than the official number of Americans missing in Cambodia.) Because the U.S. does not officially recognize the Cambodian government, it has refused to respond directly to the Cambodians regarding the remains. Cambodia, wishing a direct acknowledgment from the U.S. Government, still holds the remains. 38 Americans were killed - the last U.S. casualties in a war not quite over. A of search experts have returned to Koh Tang, hoping to recover the remains of 18 Americans left behind in the bloody battle.

The operation, which began this week and is expected to last a month, is part of the Joint Task Force-Full Accounting program set up by the Bush administration in 1992 to find the remains of more than 2,200 pilots and other military personnel missing and presumed dead.

At Koh Tang, the **missing include 10 Marines**, two Navy medical corpsmen and an Air Force pilot killed when his helicopter was shot down just offshore, and five others on the island itself.

[6] Mark Bowden, *Blackhawk down!: A story of modern war* (Penguin Putnam books, 1999 and 2000)

http://www.amazon.com/exec/obidos/search-handle-form/ 103-6593991-3811860

Web site based on newspaper series

http://www.philly.com/packages/somalia/ask/ask12.asp

[7] TEXCOM official test results of *Wiesel* air-delivery techniques

[8] Personal experience of Michael Sparks at JRTC with German Airborne units equipped with *Wiesels*.

[9] Personal interviews with Soldiers who have driven the *Wiesel* at JRTC

[10] Mak information brochure

[11] Mak LVAD test video footage provided by Bob Novogratz of RheinTech, Arlington, Virginia, December 1999.

[12] U.S. Army and Air Force M551 AR/AAV Airdrop manual; FIELD MANUAL NO. 10-515 TECHNICAL ORDER No. 13C7-10-181 HEADQUARTERS DEPARTMENT OF THE ARMY DEPARTMENT OF THE AIR FORCE, Washington, DC, 24 September 1984

http://155.217.58.58/cgi-bin/atdl.dll/fm/10-515/default.htm

[13] **GPADS** is the family of Pioneer Aerospace's state-of-the-art parafoil systems designed to use the *Global Positioning System (GPS)* for precise, autonomous delivery of loads ranging from 700 to 42,000 pounds (300 kilograms to 19 metric tons), GPADS has been demonstrated in three sizes — *Light, Medium and Heavy.* An exclusive feature of all GPADS is a Pioneer Aerospace high-glide parafoil providing up to *20 kilometer stand-off* capability and soft, into-the-wind flared landing. **GPADS-Light** has been *type-qualified by the U.S. Army* for loads in the 700 to 1,500 pound range (320 to 680 kilograms). **GPADS-Medium and -Heavy** have been demonstrated under the U.S. Army's Advanced Precision Aerial Delivery System — APADS — program. Flying beneath a 7,350 square foot (680 square meter) parafoil — *the largest parafoil in the world* — GPADS has demonstrated the soft recovery of a 36,000 pound (16 metric ton) payload — *the heaviest payload ever recovered on a single parachute.* GPADS-Light flies itself to within 100 meters (CEP) of its designated target using an autonomous navigation, guidance and control system. Pioneer's patented Mid-Span Reefing Technique ensures controlled canopy inflation and reduced 'g' loads. Pioneer is the only company to have successfully flown parafoils this large, leading NASA to choose Pioneer's GPADS-Heavy as the baseline recovery system for the International Space Station's X-38 Crew Return Vehicle. The Army-qualified GPADS-Light can penetrate winds of over 30 knots (15 meters per second) flying at a glide ratio of 4-to-1. GPADS-Medium and -Heavy demonstrated wind penetration of over 50 knots (25 meters per second) at a 2_-to-1 glide ratio.

http://www.pioneeraero.com/gpads.htm
http://www.istaero.com/gpads-m-h.htm

[14] M113 LVAD manual C1, FM 10-567/TO 13C7-16-171
http://155.217.58.58/cgi-bin/atdl.dll/fm/10-567/default.htm

[15] 5- ton LVAD rigging manual, AIRDROP OF SUPPLIES AND EQUIPMENT: RIGGING 5-TON TRUCKS, 02 MAY 1985 FIELD MANUAL NO. 10-526 TECHNICAL ORDER No. 13C7-2-481 HEADQUARTERS DEPARTMENT OF THE ARMY DEPARTMENT OF THE AIR FORCE Washington, DC, 2 May 1985

http://155.217.58.58/cgi-bin/atdl.dll/fm/10-526/default.htm

[16] U.S. Army, FM 7-7 *Mechanized Infantry Platoon and Squad*, Washington D.C.: GPO

[17] U.S. Army, FM 17-18 *Light Armor Operations*, Washington D.C.: GPO

http://www.adtdl.army.mil/cgi-bin/atdl.dll/fm/17-18/f1718.htm

[18] Fort Irwin, National Training Center experiences of the book's authors, particularly Michael Sparks who was in a M113A3 Mechanized Infantry unit supplied with M113A2 OPFOR vis-mods for the rotation.

[19] Richard Simpkin, *Race to the Swift: Thoughts on Twenty-First Century Warfare*, (London: Brassey's Defence Publishers, 1985) pp. 104

Chapter 7

Air-Mech-Strike 21st Century Tracked and Wheeled Vehicles and C4 ISR

"While there are three levels of war, the initial FGCS (Future Ground Combat System) construct is defined around the tactical level of war so as to begin a building block approach for future unit designs be they platoons, companies, battalions, brigades, squadrons or some new units that may variously be constructed, such as a cohort, a legion or a phalanx..."
—From the U.S. Army 18 January 2000 draft solicitation for the new Future Ground Combat System (FGCS)

Fort Lewis Interim Brigades: Moving out, but not Far Enough?

The Chief of Staff of the Army at the ground-breaking AUSA meeting in October 1999 stated:

> Today, 90% of our lift requirement is composed of our logistical tail. We are going to attack that condition both through discipline and a systems approach to equipment design. We are looking for future systems which can be strategically deployed by C-17, but also able **to fit a C-130-like profile for tactical intra-theater lift**. We will look for log support reductions by seeking common platform/common chassis/standard caliber designs by which to reduce our stockpile of repair parts. We will prioritize solutions which optimize smaller, lighter, more lethal, yet more reliable, fuel efficient, more survivable solutions. We will seek technological solutions to our current dilemmas.

135

The CSA's vision seeks an improved fixed-wing deployment capability using armored vehicles sized to be transportable 4-6 at a time, in USAF C-17s to deploy strategically from CONUS, and at least one-at-a-time, in-theater using USAF C-130s. In the year 2001 budget, the U.S. Congress has allocated the Army money to buy one Brigade's worth of new purchase armored vehicles but directed the Army to examine "existing equipment" to see if it will meet the Medium Brigade's requirement. Should we buy expensive wheeled armored cars that are too heavy to fly by Army helicopters, or can we build units that are both U.S. Army helicopter transportable, as well as USAF fixed-wing deployable, using existing under 11-ton M113A3s?. Does that extra up to 9 tons of armored car weight that prohibits it from flying by Army helicopters translate into more armor protection and firepower or is it lost to engines, transmissions and extra volume needed for wheels to travel/turn? Does extra armored car weight result in more or less off-road mobility, and has the recent advancement of continuous "band track" technology used in track outweigh wheeled vehicle advantages?

Vehicles Must be Sized to Fly by Army Rotary-Wing Aircraft to Enable 3-D Maneuver

The U.S. Army must carefully evaluate the current effort to field an interim, medium armored vehicle to ensure that the opportunity to gain FULL fixed and rotary-wing aircraft deployability and decisive 3-D ground maneuver is not lost over the next 10-20 years. The threat has increased in severity with the explosion of information and smart weapons technology available to the highest bidder. A U.S. Army stuck in just 2-Dimensions will find it difficult or impossible to pick or fight its way through defenses since the enemy can anticipate these courses of action.

There are a wide range of off-the-shelf, existing vehicle options and associated equipment available to gain 3-D Air-Mech capabilities today for the U.S. Army. A decision to add a few tons, going with heavier vehicles can result in an un-necessary exclusion of helo-transportability, restricting the Army to just two-dimensional maneuver in an increasingly dangerous world. A heavier vehicle does not translate into a more capable vehicle if this weight is sunk into less compact and efficient wheeled propulsion means.

As we examine these aspects in detail, it will become apparent that the majority of the Army's fighting vehicles should be tracked for maximum pound-for-pound efficiency. Being under 11-tons they are CH-47F helicopter transportable for 3-D maneuver. Tracks have the lowest possible ground pressure for maximum 2-D cross-country maneuverability. The tracked M113A3 already in Army service and the German *Wiesel* already trialed and approved for Army use can be exploited to gain decisive capabilities at the lowest cost. The *Wiesel 2*, renamed the *Ridgway* Fighting Vehicle (RFV), already uses band-tracks for propulsion, rendering wheeled-vehicle-like speeds on roads, quietness and reduced friction resulting in an amazing 7-13 miles per gallon fuel economy. The M113A3, renamed the Gavin Fighting Vehicle (GFV), lightened up by using band tracks and lightweight hatches/ramp (*Air-Mech Configuration 1*) would also enjoy the same increase in speed and reduced friction costs. Its generally understood tracks out-perform wheels off-road, and wheels have a slight edge on roads, but these conclusions are drawn from metal tracks. If band-tracks are used on tracked vehicles under 20 tons, the wheeled vehicle advantage on roads is negligible. The small band-tracked RFV of under 4-tons recommended in the Army's Recon and Air Assault units, can be transported to a conflict 2-3 at a time in C-130s, 10-12 in C-17s and unlike any other armored vehicle—by Army lease or Civilian Reserve Air Fleet (CRAF) cargo 747s in an "Airborne pre-positioning" concept.

The British Army Experience with Choosing the Right Armored Vehicle

In the late 1960s and early 1970s, the British Army wrestled with a similar new combat vehicle acquisition challenge currently facing the U.S. Army today. Their objective was to field a family of armored reconnaissance vehicles that could be utilized in a rapid deployment role projected by the Royal Air Forces' newly purchased C-130 *Hercules* cargo aircraft. The debate centered around the target weight of the vehicle. **Wheeled armored vehicles were quickly eliminated as an option due to their 28% heavier weight per square foot of troop volume gained compared to tracks.** The early-model C-130 permitted a weight of about 15-18 tons or 30-36,000 pounds (today's versions can carry 42,000 pounds). The two competing "*schools of thought*" were to either build a 15-ton plus or an 8-ton tracked vehicle. Proponents of the larger vehicle pointed to the advantages of being able to mount larger anti-tank weapons and incorporate thicker armor. The 8-ton advocates pointed out that at the lighter weight option permitted two vehicles transported for every C-130 sortie vice only one for the heavier model. A key factor in their decision was the additional advantage of the lighter option in its ability to be sling-loaded by the newly ordered CH-47 *Chinook* helicopter. A 15-ton vehicle would be significantly outside the *Chinook*'s lift envelope, as it still is today.

The British Army opted for the lighter option and fielded the 8-ton *Spartan-Scorpion-Scimitar-Striker* family of tracked armored fighting vehicles. They accepted risk in the slightly thinner armor and smaller weapon payload but gained BOTH fixed-wing and rotary-wing *Air*-Mechanized capability resulting in greater strategic, operational, tactical deployability and employability. As technology improves armor and weaponry, their overall weight decreases and compensates for this trade-off. This paid great dividends starting in August 1974, when *Scorpions* were *Air*-Meched by C-130s to secure the British base at *Dhekelia* on Cyprus after the Turkish Airborne -led invasion of the island. In the Falklands Islands war of 1982, these light ground-pressure vehicles were able to traverse the soft terrain and give battle-winning armored fire support. In the Gulf War, *Swingfire* ATGM-equipped *Strikers* boldly maneuvered forward into the Iraq without fear of being bogged down in marshy areas and successfully engaged and destroyed Iraqi T-55 Main Battle Tanks at standoff ranges. This family of AFVs then went to the mountains of Bosnia where their agility made them very helpful in escorting convoys safely along steep passes. In June of 1999, following the NATO air campaign against Serb Forces in Kosovo, the British Army used their CH-47 *Chinook* helicopters to sling armored *Scimitars* and *Spartan* personnel carriers over the mountainous terrain and the heavily mined road network to quickly gain positional advantage in their designated sector. This *Air*-Mech force had greater ground mobility, armor protection and firepower than a traditional dismounted Air Assault infantry force. This capability allowed Royal Engineers to clear the roads at a safer, less time-sensitive pace to facilitate follow-on heavier armored forces. The U.S. ground force had no such capability. The vehicle acquisition decision back in the early 1970s, had far-reaching maneuver warfare implications for the British Army. The same is true for the current U.S. Army combat vehicle acquisition effort.

Survivability *Air*-Mech Tracks or Wheels?

Paul Hornback, who is a general engineer with the federal government. presently assigned to the HQ TRADOC Combat Development Engineering Division, Fort Knox Field Office, which provides reliability, maintainability and systems engineer-

Air-Mech Vehicle superior 2-D agility (Major Charles Jarnot)

ing support to the Directorate of Force Development, Fort Knox, Kentucky, in his definitive study on, "The wheel versus track dilemma" (*http://knox-www.army.mil/center/dfd/wvtart.htm*) states:

> *With the development of any new Army combat vehicle, the question "which is better: a wheeled vehicle or a tracked vehicle?" surfaces again and again. In order to answer this question, the U.S. Army has tested and studied the merits and shortfalls of wheeled and tracked combat platforms for the past thirty years. The results of this intensive research effort indicate that no single criterion can be applied which will answer the wheeled versus track issue for all situations and missions. In fact, the underlying premise in resolving the wheeled versus track dilemma is deeply rooted in the complex variables regarding the platform's combat mission, terrain profile and specific vehicular characteristics. Tests and studies, however, established a set of criteria to determine a platform's optimal configuration.*

The 10-18 ton "*Medium*" weight armored vehicle is light enough for one to be carried in a C-130 and airlanded at an austere dirt runway without refueling facilities. If vehicle weight exceeds 18 tons, the C-130 can airdrop but not airland the vehicle, though there is no payload left to transport an infantry squad or crew on the same aircraft. Vehicles under 20 tons can be airdropped or airlanded 4-6 at a time from a C-17. The C-17 has turbofan engines which can be damaged by foreign objects (FOD) if ingested if it lands on an unpacked, dirt runway. The turboprop C-130 which is less affected by FOD is the preferred fixed-wing aircraft to deliver Army vehicles into austere runways and is the yardstick for Army Medium vehicle weight. Since wheeled armored cars are heavier for a given volume than tracked armored vehicles; the 14-18 ton armored car is 4-10 tons heavier than an equivalent volume tracked vehicle just to carry its engines, suspension, dive trains and the resultant larger armored box. **This extra 4-10 tons of** weight in an armored car does not pay-off in the form of greater armor protection or firepower over an 11-ton tracked armored fighting vehicle. Those that compare "*tracks to wheels*" citing the wheels as being heavier as some kind of "*virtue*" are "*comparing apples to oranges*" and are actually calling attention to a *weakness* in armored car design, the extra weight is swallowed up in propulsion machinery not extra armor thickness. U.S. Army Armor expert Paul Hornbeck continues:

> *A combat platform's survivability is dependent on numerous criteria to include mine and ballistic protection, size/silhouette and stealthiness. Tracked vehicles, by design, are inherently more compact than wheeled vehicles. The primary reasons for a tracked vehicle's compactness are reduced suspension clearance, wheel turning clearance, and the absence of multiple transfer cases and drive shafts that are integral to the design of multi-wheeled vehicles. Army studies have indicated that, for a comparable VCI (or ground pressure) at the same gross vehicle weight,* **wheeled platforms require up to six times more volume for drive train and suspension components than tracked platforms**. *This results in up to a 28 percent increase in vehicle volume if the same interior volume is maintained. Survivability analyses clearly indicate that* **a larger size is more readily seen and subsequently hit and destroyed.** *Additionally, as a combat platform's size increases, so does the gross vehicle weight (provided the same ballistic and mine protection are maintained), which tends to degrade vehicle mobility and deployability. The soil strength, coupled with the vehicle's characteristic ground pressure, determine a parameter entitled Vehicle Cone Index (VCI), which is a key first-order discriminator of a platform's mobility. The higher the VCI or ground pressure, the less mobile the platform becomes. Figure 1 shows that, as ground pressure increases, so does the percentage of No-Go Terrain*

(terrain over which a combat platform is immobile) due to traction loss in wet, temperate areas...As a general rule of thumb, a lower VCI not only equates to better soft-soil mobility but also indicates better performance on slopes, in sandy terrain, over obstacles/gap crossings and when overriding vegetation. From a mobility perspective, tracked vehicles offer the best solution for a versatile platform that is required to operate over diverse terrain, including extremely difficult ground, because tracks inherently provide a greater surface area than wheels, resulting in a lower VCI. Recent operations in Bosnia have demonstrated the inherent weaknesses of wheeled vehicles with regard to mobility and protection. When operations were conducted on roads, wheeled vehicles demonstrated excellent mobility and speed; but when off road usage was required and wet or snow conditions prevailed, mobility suffered.

...Consequently, Army studies indicate that **when a vehicle's mission requires off-road usage greater than 60 percent and gross vehicle weight exceed 10-tons, a tracked configuration is preferred for combat roles**. In general, wheeled platforms are more vulnerable to small-arms fire and grenade, mine, and artillery fragments, due to the inherent weakness of wheeled suspension designs, components, and tires. Wheeled vehicles may now be able to continue movement for limited distances at reduced speeds when tires are punctured by small arms rounds, battle field debris or shrapnel due to the advent of run-flat tires. Run-flat tires typically contain a hard rubber insert (some with nitrogen filled cells) inside the tire. The insert bears no vehicle load until the tire is punctured at which point, the load is transferred to the insert and vehicle movement may continue for a limited distance and speed. On the plus side, wheeled platforms provide a reduced noise signature while moving, primarily due to less vibration and metal to metal contact on running gear. Improvements in *track technology (i.e., Roller Chain Band Track) and decoupled running gear have decreased noise signatures for tracked vehicles...*

Tracked platforms do provide a skid steer capability which allows the vehicle to pivot steer (or neutral steer) and virtually pivot in place. This unique maneuver capability enhances survivability by permitting a 180 degree directional change when confined or built up areas are encountered and while traveling on narrow road surfaces. From a survivability perspective, tracked vehicles offer smaller silhouettes, reduced volume, enhanced maneuverability and better ballistic protection providing a balance that equates to a more survivable platform.

Ironically, the armored car designs the Army is looking at are all several tons over the 10 ton limit, studies have warned that is the "*point of no mobility return*" for wheels—yet some have even suggested a 20-25-ton armored car—which would be even more road-bound and unable to fly within the C-130's 42,000 pound airdrop payload limits. Ironically, an unmodified, 11-ton M113A3 hull has exactly the same armor protection as the heavier 14-16 ton LAV-type vehicle, by being a more compact size making it a smaller target than a larger armored car. But since tracked propulsion has lower ground pressure and greater cross-country mobility potential than wheels, appliqué armor could be added to 21st century tracked AFVs without mobility loss.

No Extra Airlift Required for Light Tracks: Truck Transporters/Trailers not needed

One of the misconceptions concerning wheeled armored cars is that tracks are "heavy" and "wheels" are light; such that the former needs truck transporters/trailers to move from location to location to minimize damage to roads. This drives up the total number of supporting vehicles that need to be flown from point A to B in USAF fixed-wing aircraft, slowing down deployment time and only applies to over 30-ton AFVs not under, 11-ton tracked 21st Century technology AFVs. **This constraint is**

driven by the vehicle's WEIGHT not its propulsion type. A heavy armored vehicle regardless of whether its wheeled or tracked—is going to often overwhelm third world country unimproved roads. If under 11-ton vehicles are fitted with "band track" of *Kevlar* reinforced single-piece rubber; the absolutely lowest ground pressure and soft traction possible is rendered. Scott Gourley of *Jane's Defense Review* for the week of 21 January 2000 writes about band-tracks:

> *Developed for the M113 by the U.S. Army Tank-Automotive Research, Development and Engineering Center (TARDEC) and the Canadian firm Soucy (Drummondsville, Quebec), the Band Track promises a number of advantages over conventional track systems...examples include a significant reduction in vibration, a reduction in IR signature, and a 6-7dB reduction in noise. Moreover, they note a surprising combination of traction improvement (in all conditions) and decreased rolling resistance. The decreased rolling resistance further combines with a 1200-1500 pound weight reduction (even more on the "stretched" MTVL chassis) for additional projected fuel and logistics savings...The original M113 band track design utilized rubber with aramid reinforcement. This version reportedly lasted 3000 miles in testing conducted at Yuma Proving Ground, and failed due to fatigue of the aramid. The latest version, which replaces the aramid reinforcement with steel belting, is postulated by its designers to last 6000 miles or more...Additional kit features include a sprocket and idler wear surface of molded UHMW polyethylene... Ballistic testing conducted to date has been termed successful, with rounds passing right through the material with little damage to the track....In terms of military applications, the Royal Canadian Mounted Police have recently purchased 4 sets for use on their M113s and United Defense has a set for an MTVL that will soon be installed and evaluated on a Canadian Department of National Defence (DND) vehicle. In addition, there is*

a 25-30 ton version being tested on a Bradley Fighting Vehicle (BFV) as a precursor to the Future Scout Combat System (FSCS)

Interesting note here — the Royal Canadian Mounted Police (RCMP) purchased M113s, finding with experience that they can drive over obstacles with tracks but not wheels. At the same time the Canadian Army is loaning the U.S. Army their wheeled armored cars, they are upgrading their M113A3s to MTVL standard. The stand-on-its-own, low logistics "footprint", all-terrain armored vehicle is the under 10-ton 21st Century technology tracked AFV—sized to FLY by BOTH Army and USAF aircraft!

Reduced Vehicle Weight Gains Two-for-One C-130 Split-lift *Air*-Mech Capability

When you maximize a ground vehicle for cross-country mobility, under 10-tons in weight using tracks to get the lowest ground pressure possible, you at the same time maximize its aircraft transportability in both helicopters and fixed-wing aircraft.

With the M113A3 in *Air*-Mech Configuration 1 (lightened up by using band tracks and lightweight hatches) to less than 10-tons of weight, we can enhance the deployability of *Gavins*. The M113A3 is 207 inches long (about length of a HMMWV), two are 414 inches; and the standard-length C-130H can take payloads of 480 inches long. The C-130H can carry 42,000 pounds for 1,250 miles; this means each *Gavin in AMC 1* can be combat loaded with fuel, and a driver for STOL airland operations if ground refueling is available. The C-130H can easily fly-away and airland a 11-ton GFV in combat-loaded configuration with its 11-man infantry squad, plus a 4-ton RFV and a 7-man infantry squad at the same. The latest "J" model *Hercules* with more powerful engines and propfans has even greater payload, range and speed capabilities.

Does SIZE Matter When it Comes to *Air*-Mech Aircraft Transportability?

There is a direct connection between the vehicle's propulsion system and its size/weight. Paul Hornback remarks about the impact of design size remind us:

> **Tracked vehicles, by design, are inherently more compact than wheeled vehicles.** *The primary reasons for a tracked vehicle's compactness are reduced suspension clearance, wheel turning clearance, and the absence of multiple transfer cases and drive shafts that are integral to the design of multi-wheeled vehicles. Army studies have indicated that, for a comparable VCI (or ground pressure) at the same gross vehicle weight,* **wheeled platforms require up to six times more volume for drive train and suspension components than tracked platforms.** *This results in up to a 28 percent increase in vehicle volume if the same interior volume is maintained. Survivability analyses clearly indicate that* **a larger size is more readily seen and subsequently hit and destroyed.** *Additionally, as a combat platform's size increases, so does the gross vehicle weight (provided the same ballistic and mine protection are maintained), which tends to degrade vehicle mobility and deployability.*

For the wheeled armored car/truck family to be C-130 transportable it must sacrifice firepower/armament via micro-sized turrets and may be unable to force-entry airdrop unless with the turret removed. Wheeled armored cars/trucks have axle pressures too high to be transported in cargo 747s. One technique is to use micro-sized rubber tires on armored cars to reduce their profile to more easily fit inside C-130s. The drawbacks to this imitation of your family car's "limited use spare tire" is that after you roll off the aircraft, you are stuck with narrow-tires. Another idea tried was deflating the tires for air delivery, the drawback is that it takes time; critical in STOL airland operations where vehicles must off-load and move away with the aircraft engines running, so the plane can turn around and take-off since other aircraft are usually waiting to land in the traffic pattern above.

Even the wheeled HMMWV has great difficulty due to its width, fitting inside CH-47 *Chinook* helicopters which have to have seats removed so the two vehicles can be gingerly backed in. In contrast, a tracked armored vehicle like the *Ridgway* based on the German *Wiesel* is compact enough to easily fit inside the CH-47 and weighs less than an armored HMMWV!

Mobility and Survivability: Who Can Continue to Maneuver in Combat?

Mark Bowden in his definitive work on the October 3, 1993 raid into Somalia, "*Blackhawk Down*" reports on page 271; "*...The Paki tanks would lead the convoy out into the city. Behind them, each platoon would have four [WHEELED] APCs interspersed with trucks and Humvees...because the tanks were needed to plow through the formidable barricades (ditches, abandoned shells of cars and trucks, heaps of stone, burning tires and debris) the Somalis had erected to block most of the main roads leading out of the UN facilities...*"

B. H. Liddell Hart in *The German Generals Talk*, Chapter XIII, Frustration At Moscow, Pg. 167-179 states:

> *Here is the most startling of all. What saved Russia above all was not her modern progress, but her backwardness. If the Soviet regime had given her a road system comparable to that of western countries, she would probably have been overrun in quick time, the German mechanized forces were baulked by the badness of her roads.*
>
> *But this conclusion has a converse. The Germans lost the chance of victory because they had based their mobility on wheels instead of on tracks. On these mud-roads the wheeled transport was bogged when the tanks could move on. Panzer forces with tracked transport might have overrun Russian's vital centers long before the autumn, despite the bad roads. World War I had shown this need to anyone who used his*

eyes and imagination... Such country was bad enough for the tanks, but worse still for the transport accompanying them, carrying their fuel, their supplies, and all the auxiliary troops they needed. Nearly all this transport consisted of wheeled vehicles which could not move off the roads, nor move on it if the sand turned to mud.

Sir Archibald Wavell, one of the British Army's few Field Marshalls (equivalent to our 5-Star Generals) said in WWII "*There is no room in war for delicate machinery*". Under 11-ton tracked AFVs have demonstrated greater off-road mobility/reliability than wheeled armored cars/trucks. Tracked propulsion offers larger ground traction, a more compact arrangement making the vehicle lower for aircraft internal carrying and an ability to rumble over obstacles. Paul Hornback writes:

> As off road usage dominates the vehicle's profile, tracked configurations provide significantly better mission travel times. Consequently, Army studies indicate that when a vehicle's mission requires off road usage greater than 60% and gross vehicle weight exceeds 10 tons, a tracked configuration is preferred for combat roles. However, when the gross vehicle weight exceeds 20 tons and off road usage remains above 60%, a tracked configuration is required to guarantee the best mobility for unrestricted all weather tactical operations.

Available data proves the M113A3 *Gavin*-MAV and/or M8 AGS tracked AFVs are more survivable than large wheeled vehicles. This is because they can force their way through and over obstacles and use add-on modular armor fitted, which as technology improves, so does armor protection, without loss of mobility due to the extra power/traction available via the efficiency of tracked propulsion means. *Gavins* can push mine rollers/plows to prevent mine run-over incidents. *Gavins* that run over mines can be run on "short tracks" avoiding the roadwheels damaged in the blast. *Gavin*-MAVs with lighter "band tracks" are resistant to small arms fire that would deflate and shred armored car tires that need shaping to roll. The BCT can resupply itself x-country with designated *Gavin*-MAVs to avoid road MSRs entirely; an asymmetric "paradigm buster". The small size of Ridgway Fighting Vehicles make them difficult to detect and hit. Their steel armor is proof against 7.62mm ball small arms, the same basic level as the much heavier LAV-25.

Appliqué' armor is rarely if ever actually added onto armored cars because it would smother what little cross-country mobility they have. The armored car with a large volume to carry a 9-man infantry squad, and have clearance for its wheels, ends up being a very huge silhouette, easily targeted and hit by the enemy. Armored cars are often heavier than tracked AFVs, but this isn't in the form of thicker armor protection, but weight for the vehicle's construction. The only thing an armored car can do well in terms of survivability is take a mine blast by sacrificing the wheel/axle.

Infantry Carrying Capability

Jose Vilabla, Spanish General and historian once said "*The Soldier is the primary and most powerful mechanism of war*". The Army's move to an infantry-centric force prepares us for the spectrum of conflict. However, what is the optimum vehicle to move infantry in?

On the ground, at first glance, large AFVs dominate over small AFVs in that they have more room to carry infantry, however; if in the constraints of a C-17 you can get more small AFVs in the space that large AFVs can be carried, the small AFV may actually be a better force structure choice. The efficiency is how many infantrymen can be delivered per vehicle volume and the ability to deliver more combat power per C-17 aircraft in a strategic *Air*-Mech movement.

The *Ridgway* which can carry 7 men inside yet takes up half the space of a large AFV like the LAV. By fitting 12 *Ridgways* per C-17 you get 84 men mobile in AFVs. *Gavins* can carry 13-infantrymen. Only the *Gavin* (MTVL) stretch version has the volume to carry a full 9-man infantry squad AND a 25mm autocannon BFV turret or 9 dismounting infantry. A M113A3 with a smaller one-man 25mm/7.62mm turret or our proposed 30mm M230 autocannon/*Javelin* ATGM cupola can do this, also. Both configurations have 2 Soldiers as vehicle crew. In the case of the LAV-25, only 4 dismounting infantrymen can be carried with a 25mm turret.

The *Ridgway* can carry 7 men. Upon dismounting, just the driver. or the Driver and Gunner can remain, giving at least a 5-man infantry element. Two *Ridgways* can carry a full 9-man rifle squad.

Supportability: Which Costs Less to Operate?

A combat platform's supportability is dependent on numerous factors to include fuel usage, reliability, and O&S costs. Wheeled vehicles until recently offered better fuel economy due to the reduced friction losses inherent in wheel/tire suspensions and running gear, though with band tracks this is no longer true. Better fuel economy translates into smaller on-board fuel storage requirements or greater operating ranges for platforms.

A recent February 28, 2000 report by Kim Burger from Inside the Army shows that "A closer review of data shows wheels not more reliable than tracks":

> *Further analysis of past studies on the characteristics of wheeled and tracked vehicles has revealed that wheeled vehicles are not considerably more reliable than their tracked counterparts, discrediting yet another widely held perception about the advantages of wheels, the technology director for the Tank-Automotive and Armaments Command told Inside the Army. The finding lends further weight to earlier Army findings that wheeled platforms do not cost significantly less to operate than tracked platforms (ITA, Dec. 27, 1999, p1).*
>
> *...wheels initially appeared to have better reliability ratings, a difference that should have been reflected in lower operations and sustainment costs. However, the costs were coming in at about the same level. A*

source told ITA then that the Army would continue evaluating the data to understand the "discrepancy". [TACOM Representative] McClelland said evaluators looked closely at the raw data from the past tests and spoke to some of the people involved in the scoring conferences. They found that most of the wheeled vehicle tests were conducted on roads, which would likely give them better reliability results than if they had been driven on rougher terrain, he said. Tracks were generally tested on more challenging terrain. The data says, yes, the wheels have better reliability....[However] tests were not run on the same courses at the same time, McClelland said. After scrutiny of the data, it does appear those numbers have come substantially closer, and, as a result, explain why the cost data, was so similar... The finding also illustrates why old reliability data is not valid for use in current acquisitions, and why the Army would run new tests in which all vehicles run the same course, with the same scoring conference, to evaluate proposals, said McClelland. Reliability is an important factor because it affects the amount of spare parts, and the cost of those parts, which are principal O&S cost drivers. **Analysts have also found today's wheeled vehicles typically are outfitted with complicated drive trains that can further affect their maintenance score**, McClelland said, if the Army ran a competition that involved removing and changing power packs, many tracked vehicles — such as the Armored Gun System, the Abrams tank and, eventually, the Crusader artillery system — would probably win hands-down... Wheels also are not necessarily easier to repair than tracks. The Army wants a vehicle that can run on flats. But when it comes to replacing the tires, they are difficult at best and dangerous to repair if not done on a proper machine, McClelland said.

Paul Hornback adds:
However, one must bear in mind that wheeled ve-

hicles generally have a higher percentage of on road usage while tracked vehicles incur more off road usage. Obviously, the more severe cross-country terrain results in reduced reliability for the tracked vehicle. A recent test of the Up-Armored HMMWV running a scout profile with 68% off road travel resulted in significantly lower reliability when compared to the same platform running at a tactical truck profile of only 40% off road.

Cargo 747 Airliner *Air*-Mech Transport of RFVs and Army Helicopters

Maurice de Saxe, 18th Century Marshall-General of France concluded that: *"It is not the big armies that win battles, it is the good ones".* A good Army has multiple options to achieve battlefield victory. Small sized tracked AFVs are 3-D maneuver transportable in both fixed and rotary-wing types of U.S. military aircraft and are more 2-D cross-country, terrain mobile, while having additional protection options, and carry more infantrymen.

A key feature of the 3-4-ton RFV equipped RSTA squadron in each *Air*-Mech-Strike BCT is that this force can be moved by leased or CRAF cargo 747s, freeing up USAF aircraft to transport oversized vehicles. The 3-4-ton RFV on band-tracks (low ground pressure) is gentle enough on the floors of cargo 747s so that 25 can be deployed to an ISB. Three UH-60L *Blackhawk* helicopters or 8 x OH-58D *Kiowa Warriors* can be flown inside a 747s by virtue of their weight being spread over a large area. Once in theater, UH-60L helicopters can fly the RSTA vehicles over the battlefield.

A new *Air*-Mech Brigade force structure with Recon RSTA RFVs would dramatically improve their rapid deployment via Civilian Reserve Air Fleet (CRAF) or long-term lease Airborne pre-positioning. Long distance sustainment, peacekeeping and warfighting capabilities would be exponentially increased by freeing USAF aircraft to deliver oversized battlefield operating systems that otherwise are unavailable within 72 hour time constraints. This strategically-agile Medium Force could operate across the full spectrum of conflict, in all terrain types by being a 3-Dimensional Air-Mech-Strike maneuver force. The current Army Medium Brigade plan is only focused on deployability by USAF aircraft to the operational area. Airborne and Air Assault enhancements to maneuver are not being exploited.

A Future for *Air*-Mech in the U.S. Army Can be Tied to the "Interim Brigade"

The U.S. Army's Interim Brigade transformation effort "*center of gravity*" is the selection of the vehicle types. While tomorrow's Army may be equipped by some variant of the Future Combat System, the near term vehicle selection will be what is used to fight wars until 2015. In a 16 February 2000 speech before the Association of the U.S. Army winter symposium in Florida, the Army Chief of Staff described three different types of operational forces:

> *An 'objective force' that's about 8 - 10 years out; an 'interim force' that will bridge the gap and capabilities between what we have today and what we hope to achieve with the objective force; and a legacy force —today's force — that magnificent force of heavy divisions and the best light infantry in the world that we live with today. And if there is a miscalculation [by a hostile nation] in the next 8 - 10 years, maybe 15 [years], that's the force we'll go to war with.*

The draft Operational Requirements Document (ORD) for the family of Medium Armored Vehicles (MAV) that will be used to equip these Interim Brigade Combat Teams (IBCTs) describes two different phases of fielding: an "Initial" IBCT that will be equipped with off-the-shelf hardware and a follow-on IBCT that will receive off-the-shelf hardware that has been enhanced with selected technology insertion programs.

Representatives for the U.S. Army's Training and Doctrine Command (TRADOC) have publicly enunciated an expectation that the Army's interim force structure(s) will be "*dismounted infantry-centric*" forces and that the pivotal vehicle system or pacing item, will be the **infantry carrier "Interim Armored Vehicle" (IAV) or "Medium Armored Vehicle" (MAV) variant**. As the "*center of gravity*" in the IBCT transformation effort, the infantry carrier's capabilities and limitations will drive all other tactical, operational and logistical planning.

We suggest that the Army choose a progression of MAVs that support 3-D maneuver with Army helicopters, and USAF fixed-wing aircraft., creating an Interim Phalanx Force (IPF) capability for the Army transformation in the 2001-2015 time frame. Additionally, the MAV selected must be fully 2-D maneuver capable once on the ground, with mobility which includes moving over obstacles, operating across country and provides the best armored protection available. Consequently, IBCT/IPF medium armored vehicle selection may be the Army's one and only "*move*" prior to any combat situation that may arise in the 2005-2015 time frame.

Early Army documentation reflects the notional IBCT/IPF to be a 3500-man unit equipped with approximately 331-400 AFVs. The Army is conducting a solicitation/procurement for these AFVs under the designation Family of Medium Armored Vehicles (F-MAVs). While some see the procurement as simply a matter of "Tracks vs. Wheels," in reality it is much more complex — with additional considerations like 3-D helicopter transportability, strategic fixed-wing aircraft mobility, tactical 2-D mobility (both on-road and off-road) crew survivability, and lethality.

Could Existing Systems be Enhanced for a Better, *Air-*Mech-Strike Capable BCT?

Technology insertions could be applied to today's "legacy" fleet which includes TRACKED MAVs with superior performance that would enhance strategic mobility, higher survivability, greater tactical mobility, and a more lethal Interim Phalanx Force—capable of *Air-*Mech-Strike maneuvers at a drastically reduced cost. Cost savings could be applied to the acquisition of a number of complementary systems that provide the BCT/IPF with additional capability.

With a stated goal of converting five BCTs to the "Interim design", at a cost of several **billion** dollars, why not get an additional capability? A mix of already-in-Army-service M113A3 "*Gavin*" Fighting Vehicles (GFVs) and a purchase of the smaller *Ridgway* Fighting Vehicle (RFV) based on the German *Wiesel Air*-Mech Vehicle, provides that cost-effective additional capability. The GFV would be para-mech capable, helo-transportable with the fielding of the new CH-47F model *Chinook* and provide a credible armored personnel carrier. The RFV would be cargo 747 strategic transportable, helo-transportable via the current fleet of UH-60L *Blackhawk* helicopters and provide an excellent platform for the Reconnaissance Surveillance and Target Acquisition (RSTA) Squadron, and as a fireteam infantry carrier for the 101st Airborne (Air Assault) Division.

Assuming that the Army would have to spend $800,000+ in new acquisition costs per large wheeled MAV system (a conservative estimate), and another $200,000+ for a turreted weapons, sensor or C4 ISR system, it will cost roughly $1 million per new large AFV purchased — translating to more than $400 million per BCT/IPF conversion. Our data suggests that the GFV & RFV option is a more efficient, low-cost, competitor for the Army's Medium Armored Vehicle acquisition effort. About 50% of the armored vehicles in an Army Heavy unit are ALREADY M113A3 based. Why replace them? Why not follow a conversion formula to:

1. **Build on what we have** (M113A3s) to save money by using proven combat vehicles
2. **Buy what we need** (weaponry, armor, band-tracks for M113A3s, *Ridgway Air*-Mech vehicles for RSTA and 101st Airborne Division) to attain 3-D maneuver
3. Integrate with our powerful (M1s/M2s) for 2-D maneuver

Two different M113 "*Gavin*" conversions could be undertaken; both at significantly low cost. The first involves a vehicular powerpack / performance conversion into the Army's more recently fielded M113A3 configuration — providing an armored troop carrier as fast and mobile off-road as current

"*Abrams*" and "*Bradleys*" but capable of deployable on C-130 aircraft and with sufficient horsepower-to-ton ratios to further enhance Soldier survivability with a range of add-on armor packages. The conversion approach advocated here also ensures the least turmoil in the Brigade, building on the M113A3 that already exists and an upgrade / conversion process that has already been proven at Anniston Army Depot (ANAD).

The second possible M113 upgrade not only improves vehicle performance but also features the introduction of new hull plate sections that serve to "stretch" the vehicle. The lengthened chassis and greatly increased horsepower not only permit add-on armor but also allow adaptation of 1 or 2-man turrets from the TOW / 25mm / 7.62mm type weaponry already existing for the *Bradley* Fighting Vehicle. The stretched M113A4 would most likely place the vehicle outside of the new CH-47F *Chinook* helicopter lift capability as would the mounting of an elaborate two man turret, limiting the vehicle to fixed-wing airdrop or air-land operations.

The first two "stretched" M113A4s are reportedly scheduled to enter U.S. Army inventories this year in the form of smoke generation systems. It is likely that additional integration engineering work would be required to insure C-130 transportability of the turreted *Gavin*-MAV versions, but its feasible not an impossibility like it is for most high-profile wheeled armored cars which are simply too high to fit turrets and roll under the C-130's 102" height restriction.

A viable option could be to convert some of the thousands of old M113A2s the Army owns to replace the BFVs and M1 *Abrams* tanks in selected heavy Brigade battalions — with the projected ability to convert approximately 8 x M113A3s for the price of a single new wheeled armored car. Moreover, M113A3 logistics are already in the Army inventory while the introduction of any new system will saddle service support agencies with an array of additional expenses.

Under this plan, M113A3 *Gavins* would be modified for a family of variants, infantry, anti-tank, mortar, with a modest purchase of small RFVs for the RSTA Squadron and 101st Airborne Division. The *Gavin*-Mobile Gun System variant could be a large caliber gun turret on the M113A3 and/or a frugal buy of M8 Armored Gun Systems. Since the M113A3 *Gavin*-MAV is already in Army service, the only new acquisition would be for the RSTA's small AFV and perhaps the M8 Armored Gun System. The BCT would either be 50% *Gavin* and RFV, or 50% *Gavin, Ridgway* and M8 Armored Gun System.

Military Occupational Specialty (MOS) 19-series *Abrams* tankers could convert to the mobile gun system AFVs. MOS 11-series Soldiers who were once *Bradley* Fighting Vehicle (BFV) crewmen would transition to the possibly "stretched" and "turreted" M113A3/4 variants, some having the almost the same M242 25mm automatic cannon/7.62mm coaxial MMG type turrets. M1 tanks and BFVs exchanged for M113A3s and/or M8 AGSs and small AFVs would be transferred to a designated Enhanced Ready Brigade in the National Guard without M1s/M2s to be go-to-war, ready at their MATES/UTES sites. BFV units converting to *Gavins* will not need as many mechanics, so these personnel slots can be used to fill the RSTA Squadron.

If the new BCTs are based on the already in Army service M113A3 *Gavins*, we should be able to obtain a new AFV type. In this case, the RSTA's RFV could be a "new and different type" with capabilities not now possessed—like B-747 cargo jet and UH-60L helicopter transportability.

Air-Mech-Strike and IBCT Acquisition Costs

The current IBCT model of 400 MAVs as new purchases requires major budgetary help from Congress to effect.

Costs to Build a 400-Vehicle Brigade Combat Team with only 2-D Maneuver Capabilities

400 LAV-type 14-20 ton armored cars @ $880,000 each = $387,200,000
33 x Weapons company special turrets @ $250,000 each = $8,250,000
TOTAL: (Estimate) $395,450,000

The year 2001 DOD budget for the Army authorizes $537

million for the first IBCT, setting aside more than $1,000,000 (million) dollars for each of the 300-400 IAVs. In contrast, the newest M113A3s were built in 1992; currently A2 models are being converted at Anniston Army Depot in Alabama at a cost of about $100,000 per vehicle. Conversion has also been done in Korea, Mainz Germany and at Red River Army Depot in Texas. Force XXI Battle Command Brigade and Below (FBCB2) C4 ISR computer equipped M113A3 type AFVs are being fielded to the 4th Infantry Division (Mechanized) beginning in the year 2000.

An Army Heavy Division today (though it varies between Divisions) has 862 AFVs, 330 are M113A3 types, 207 are M1 MBTs and 196 are M2 BFVs or about 50% of all the Division's AFVs. Thus, an Army Heavy Division is already half "Medium" weight MAVs used by combat engineers, medics, and maintenance and staff headquarters units. We need to keep in the Brigade Combat Team a Battalion of M1s and a Battalion of M2s for heavy 2-D shock action, and only convert the third Infantry Battalion to M113A3 *Gavins* and build the Recon Troop into a RSTA Squadron with *Ridgways*. This means 44 BFVs will be replaced by 44 lighter *Gavins* for the Rifle Squads. So much is saved in terms of cost, weight and manning (5 *Gavins* per C-17 compared to 2 BFVs, only 2 men needed to crew a *Gavin* compared to 3 in a BFV) we add a 3 vehicle 81mm mortar section to each *Gavin* Rifle Company and a 33 *Gavin* Weapons (Mobile Gun, Anti-Tank, 120mm Gun-mortar, EOCM variants) Company to the *Gavin* Battalion.

Note: we recommend the tested and type-classified M8 Armored Gun System as the Mobile Gun System (MGS).

Cost to build 2D/3D Maneuver-Capable *Air*-Mech-Strike Brigade Combat Team

GAVIN INFANTRY BATTALION ACQUISITION COSTS
53 x 11-ton M113A2s converted to M113A3
  @ $100,000 each = $5,300,000
33 x Weapons Company *Gavin* special turrets
  @ $250,000 each = $8,250,000
TOTAL: (Estimate) $13,550,000

RIDGWAY RSTA SQUADRON ACQUISITION COSTS
40 x RFVs (*Wiesel 2s*) @ $400,000 each = $16,000,000
60 x ATVs @ $3,000 each = $180,000

———————————————————————————
$16,180,000
13,550,000

GRAND TOTAL: (Estimate) $29,730,000

400 LAV-type BCT wheeled armored car-from-scratch
   approach $395,450,000
86 *Gavins*, 100 *Ridgways* and ATV BCT $29,730,000
Money saved with *Air*-Mech-Strike BCT $365,720,000

Note: Allows conversion of additional Battalions in every Division of the active Army.

OPTION A:
M8 Armored Gun System option: if the 12 x MGS are the M8 with its 105mm shoot-on-the-move capability, it will cost $3,500,000 each instead of $250,000 for a MGS turret on a M113A3/4.
Grand Total cost for the *Gavin* Battalion would be $68,730,000, still $336,720,000 less costly than the wheeled armored car-from-scratch approach.

OPTION B:
If each Platoon Sergeant's GFV were equipped with a 25mm/7.62mm one-man turret at $250,000 each, 9 x $250,000 = $ 2,250,000 for a total cost of $31,980,000 for the *Gavin* Battalion, still $363,470,000 less costly than the wheeled armored car-from-scratch approach.

Costs to build multiple BCTs within the 10-Division Army
8 BCTs @ $395,450,000 each = $ 3,163,600,000 or $3.163 BILLION DOLLARS
Combat power = a 8 BCTs available for missions, rest of the Army is unchanged!

To convert 8 Brigades from scratch to a 400-MAV force requires changing every Battalion in these Brigades to a new type MAV. If this includes 3 Rifle Battalions, 1 RSTA Squadron, an Artillery Battery and a headquarters, it translates into 6 Battalion equivalents per Brigade for a total of 48 Battalions being changed. In contrast, the *Air*-Mech-Strike concept can change almost double the number of Battalions to a higher level of 2-D/3-D combat capability at a lower cost.

30 *Air*-Mech-Strike BCTs @ $29, 730, 000 = $891, 900, 000 MILLION
Combat power = 30 BCTs, ENTIRE U. S. Army transformed into 2D/3D maneuver capable, mission ready status

Option A: with 360 x M8 Armored Gun Systems = **$1,260,000 BILLION**

Option B: with 810 x PSG GFVs with 25mm/7.62mm turrets = **$959,000,000 MILLION**

Option C: 1 Battalion in the 1st BDE of 101st Air Assault Division with 104 RFVs instead of 86 GFVs $ 47, 780, 000 MILLION for total of **$909,950,000** for all 30 BCTs

Option D: 2 Battalions in each of the 4 Light, Airborne, Air Assault Division's BCTs receive a Combat Support Vehicle large ATV in every squad to upgrade mobility = 664 ATVs @ $100,000 each = $66,400, 000 MILLION, total AMS conversion: $ 1,555, 900,000 BILLION for all 30 BCTs

The *Air*-Mech-Strike model of changing one combat battalion and creating a RSTA squadron in every brigade provides beneficial mobility capabilities to an already lethal force. This equals 60 total Battalions being changed to new vehicles close to the same time frame at 1/10th to 1/3 of the cost of the from-scratch IBCTs. It is the best course of action for the U.S. Army providing more flexibility and capability for its money with the *Air*-Mech-Strike approach. For the cost of about two 400-MAV BCTs, we can outfit the ENTIRE active Army—30 BCTs to become 2D/3D *Air*-Mech-Strike capable!

At the same time our reserve component receives additional M1/M2 fielding. Army funding could be used for "*tech insertions*" of weaponry, stealth materials, sensors, C4 ISR and urban war attachments like forklift or boom arm capsules to prevail in urban conflicts.

Air-Mech-Strike Reconnaissance and Logistics: Wheels up, Scouts out!

Field Marshall Napoleon once concluded: *"You can never have enough reconnaissance"* To gain that reconnaissance, in *Air*-Mech-Strike we exploit armored tracks at their optimal weight and configuration and advocate the same for unarmored wheels, making sure both are sized for *Air*-Mech delivery by fixed and rotary-wing aircraft.

"*Wheels up*" is a term often used by military planners and operators, but it's especially critical in regards to force projection and getting to the fight and/or conflict rapidly. The *Air*-Mech-Strike 3-D maneuver concept demands careful analysis and exploitation of wheeled platforms, primarily for reconnaissance, command and control and logistics.

Small-wheeled ATVs can be easily silenced and/or provided hybrid diesel-electric drive systems compared to large wheeled armored cars. There is a role that these light wheeled All-Terrain Vehicles (ATVs) and human-powered All Terrain Bikes (ATBs) and carts (ATACs) can operate in the 3-D *Air*-Mech-Strike force because they are quieter than many kind of AFVs and still be terrain agile. This role is to be the "feelers", scouting ahead of the *Air*-Mech-Strike force on the ground to gain the observation/firing position required without alerting the enemy.

The second role is Load Bearing Vehicle (LBV) or "Combat Support Vehicle" for the Soldier's rucksack so he can move without the energy expenditure, as well as move him and his fire-team if necessary. The model here is of *Merrill's Marauders* in WWII China-Burma-India operating behind enemy lines using "*pack mules*" to carry supplies/equipment, except this mule is mechanical.

All-Terrain All-Purpose Carts (ATAC)

Small wheeled Human Powered Vehicles (HPVs) ATBs or ATACs and motorized ATVs can sublimate themselves by folding, becoming a backpack, being carried on another vehicle, on a trailer or stacked on pallets or para-dropped inside Container Delivery System A-21/22 containers to save space for air transport. These vehicles do not suck up fuel at a huge rate, and in the case of HPVs, fuel costs are ZERO. A portion of the *Air*-Mech-Strike force needs to be extremely terrain agile on foot and/or with HPVs; if this part gets wedded to a large motor vehicle, they will simply not spend enough time on physical conditioning to fight this way. Parts of the *Air*-Mech-Strike force must be able to fight on foot initially to force/secure an entry DZ or LZ, then maneuver through closed terrain on foot, aided by HPVs or hitching a ride on the back of M8 AGS light tanks or transport helicopters. ATACs like the Ferno UT 2000 system already in use by a dozen armies around the world could have a kevlar lining to protect from small arms fire as expedient frontal cover be a fighting position roof for artillery/mortar burst overhead cover and could be fitted with provisions to be towed by ATVs and larger military vehicles. We advocate fielding an ATAC like the UT 2000 as a Split Load Trailer (SLT) to carry Soldier rucksacks and ammunition for *Gavins* and *Ridgways* when the conditions arise that weight has to be transferred for split-lift *Air*-Mech movements. The feature of the SLT would be its ability to be towed by hand like an ATAC but can be hitched to the *Gavin* or *Ridgway*'s towing eye and be vehicle towed.

All/Extreme Terrain Bikes (A/ETBS)

One way to create a fully 3-D *Air*-Mech-Strike recon capability is by creating "*Dragoon*" units: heavily armed mounted Soldiers who use diesel-powered All-Terrain Vehicles and folding bikes and carts. "Dragoon" Light Bicycle Infantry (LBI) can use air-droppable folding ATBs and Extreme Terrain Bikes (A/ETBs) with 10-inch wide tires to traverse over sand/snow...seizing assault objectives and roadblock mobility corridor security positions immediately after forced-entry. The maneuver of LBI units would be supported by Dragoons in ATVs firing heavy-caliber machine guns, grenade launchers, rockets and recoilless rifles. Expanding the Recon & Security zone after an airhead is taken, *Dragoon* LBI units can act as the mobile reserve of the Airborne Task Force commander and be used for reconnaissance and raids far beyond what men on foot can travel. The Airborne's battle space tied in with digital communications means extends beyond enemy artillery and rocket weapons ranges and secures the foothold for offensive operations to collapse the enemy before he can react.

HPVs can be jumped over the rear ramp of high performance aircraft like the Lockheed Martin C-130 *Hercules* or the super-large side jump doors of the C-17 *Globemaster III* used by the USAF. Using a padded airdrop bag for folded ATBs to be jumped with common Airborne combat equipment attached to ready-for combat individual Paratroopers as lowering line loads. Folded ATBs and ATACs can be Container Delivery System (CDS) bundles to be recovered later as the 1st Tactical Studies Group (Airborne) Paratroopers did for the Operation: DARK CLAW demonstration, February 11, 1993 at Laurinberg-Maxton airfield near Fort Bragg, NC. "Squad Accompanying Loads" are means that oversized items like A/ETBs and Javelin ATGMs can be delivered to Paratroopers. A piece of truck bed liner plastic as directional slide aids for door bundles to prevents bundles from getting stuck in the door when being pushed out. Its highly desirable to have A/ETBs and small unarmored and/or lightly armored wheeled ATVs working in concert with *Ridgway* AMVs in the RSTA troop to effect effective stealthy reconnaissance as light tracked tanks and 4x4 jeeps worked in concert with foot scouts did in WWII. Small AFVs with helicopter transportation assets in the Brigade Combat Team MTOE enable 3-D battlespace dominance.

Imagine what an *Air*-Mech-Strike force with Flyer 21 Large

ATVs and Polaris small ATVs could achieve if 3-D inserted deep into the enemy's undefended rear areas by Army aviation, with 1,000 mile ground range by carrying its own re-supply of fuel, *Javelin* or TOW ATGMs, grenade launchers, machine guns and modern high tech communications, computers, GPS, COLT Teams, Attack helicopter/Close Air Support and well trained, bold and daring leaders, scouts and anti-tank gunners? That's where "*Scouts Out*" is especially critical, the synergy of "*wheels up*" to get your "*scouts out*"!

Motorized All-Terrain Vehicles

The proper role of the "wheel" has been the subject of great controversy within military circles, particularly the relatively heavy "armored car or truck" of 14-20 tons with six to eight wheels on 3-4 axles when compared to the opposite extreme, the 70-ton M1 *Abrams fully tracked* main battle tank. There is plenty of missing ground in that comparison; the focus here is the exploitation of modern wheeled 4 x 4, off-road technology in LIGHT platforms that do not smother their off-road mobility potential, by trying to be a "wannabe" tank.

The issue of protection is where the critical analysis of task and purpose relates to the use of armor on wheeled and tracked vehicles. Too much armor will take away ALL cross-country mobility from a wheeled vehicle, and on a tracked vehicle make it require ever increasing amounts of power/fuel to propel it that could easily become too heavy in general for unimproved areas in the Third World.

The first component of protection is operational security and stealth deception so the enemy doesn't detect you to hit you. Skillful (stealthy, undetected) reconnaissance and counter-reconnaissance *aids* force protection. The most important battlefield give-away to the unaided human senses is the path predictability of the platform, followed by its noise, its visibility, with height and dust clouds as the biggest drawbacks. With sensors, the noise, heat and radar signatures of vehicles become more problematical. The inevitable conclusion is smaller and more compact is better for stealthy recon than larger and bulky. Reconnaissance precedes all successful operations.[1]

The issue of ground pressure per square inch is just as critical to a farmer as it is to a military professional, the key difference though is that somebody is searching diligently to shoot at and kill you and your military platform, not your tractor. Considering that, the issue of protection and firepower requires fighting vehicles to have various thresholds of armor protection that allow depending on the mission and an ability to carry a heavy enough weapon (cannon, large missile system, howitzer, etc.) to engage the enemy successfully within mobility mission constraints.

Combat fundamentals – firepower, protection, maneuver and leadership – are the first steps at understanding the Battlefield Operating Systems (BOSs). All armed forces strive to have a fine balance between them, as each affects the other.[2] These fundamentals apply to all platforms – air, land and sea. Leadership is the cornerstone of the four. These fundamentals are vital in the understanding of *Air*-Mech-Strike land and air platforms in both strategic force projection and operational and tactical employment.

Cargo and armor on a platform is a function of weight, and the ability of the chassis, suspension and power plant to carry and move that burden. Whether it is 20 tons of hay on a tractor-trailer or 20 tons of armor (protection) on an M1A2 tank, it's still a criteria and deals with *what* the task and purpose is. Drawing or pulling a load, whether a farm implement or howitzer behind a 5-ton capacity truck, it's also a physical consideration. The issue of firepower can be confusing particularly between heavy cannon systems (not unlike the crane on tracked construction equipment for stability and agility) and missile systems like LOSAT, *Hellfire* and TOW (comparable to lift bucket on utility truck on *wheeled* platform on a street in your town or city). Tracked vehicles still operate on "wheels" to carry forth the track that will distribute the great weight throughout the entire track. The many small "road wheels" of the track perform the task of distributing energy much like the many wheels in pulleys that lift great weights. Automotive engines deliver great torque and strength when in low gear engaging many gears (wheels). If a platform doesn't need to carry the weight of armor and large gun systems, then fewer wheels are necessary, thus reducing weight, a vital characteristic for AIR Deployability, Sustainability and certainly Agility...the other four characteristics – Lethality, Responsiveness, Survivability and versatility – all characteristics of the Army's Transformation Force Plan set forth by the Army Chief of Staff.

Firepower and maneuver, whether grand operational movements or the simplicity of "shoot, move, communicate and survive" at squad level, are the fundamentals that achieve the victory through offensive action. Protection allows friendly forces to preserve combat power to conduct fire and maneuver. The great advances in lethality through advanced missile and digital technology offers lighter forces the ability to at least meet, halt and even defeat an enemy heavy armored force. Maneuver, or mobility, is the 2-D or 3-D movement of combat forces to achieve a *positional advantage*, usually to observe/report and/or deliver direct (close) and indirect (far) fires. Light, wheeled reconnaissance and anti-tank vehicles always had a capability to place indirect standoff fires IF they can get there undetected and survive. But HOW do wheeled vehicles get in position to observe, report and even call fires in on the enemy? *Air*-Mech-Strike! offers the ability to achieve tremendous positional advantages, through a 3-Dimensional operational doctrine, and wheeled reconnaissance platforms by being the smallest and lightest ground platforms available and hence the most stealthy—are a key part of that doctrine with U.S. Army Aviation – hence, *Wheels Up, and inside a Chinook CH-47 or slung underneath a Blackhawk UH-60. Air*-Mech-Strike doesn't replace heavy or light forces, it complements *both*.

Air and Land Mobility

The merging of cars and airplanes have always captured the imagination of the public. Inventors like Robert Edison Fulton Jr. (he designed *Skyhook* Surface-To-Air (STAR) recovery system, used by Air Force MC-130 *Combat Talons* and Army SF for extracting Soldiers by fixed wing aircraft) did achieve some limited success in 1949 with the "*Airphibian*", a propeller driven automobile, in which the wings and tail could be stowed away while driving the invention as an automobile.[3] That technology is even more developed today. U.S. Army Aviation and ground reconnaissance must have the capability to maneuver rapidly and conduct deep operations by combining the capabilities of both air and ground vehicles. This includes both tracked armored fighting vehicles like the Gavin-MAV, Ridgway Wiesel, and reconnaissance, command/control and logistics light wheeled platforms such as the Flyer 21 4 x 4 utility vehicle (Flyer Group, California) and the Polaris Diesel All Terrain vehicle (ATV), which are both "High Mobility Multi-Purpose Wheeled Vehicles" by any definition of their capabilities. ATV technology of the eighties has been fully developed and employed on every continent; it is unfortunate that the technology wasn't as advanced in the eighties, when the 9th Infantry Division was the High Technology Test Bed for a "medium" type of division at Fort Lewis. Advanced designs and metallurgy offer stronger frames, components and greater durability for All-Terrain Vehicles (ATV) than what were available then. This translates today to greater speed, agility and ability to carry *heavier weapons and payloads*, while still being light enough for fixed wing and rotary aircraft for 3-D maneuver.

The True 'Iron Horse' and "Little HMMWV"

Historical references to the 'Iron Horse' refer to Native-American interpretations of the locomotive during the 19th Century and settlement of the American West. The mustang-mounted Plains Indians probably never envisioned the 'modern' 'Iron Horse' – the individual rider's ATV. Instead of four legs, 1400 pounds of bone, muscle, hide and attitude, the American farmer, rancher, hunter and recreational user now have a four wheel drive, 535-750 pound, one or two cylinder "Mount" that can also CARRY up to 270 additional pounds of tools, equipment and lethal weapons.

This single machine has created a significant paradigm shift in American agriculture's move to modern, convenient and cost effective mobility. There are more horses being ridden for pleasure and recreation in the last thirty years than actual working horses being ridden by cowboys, ranchers and farmers.[4]

Family farmers and ranchers especially took to the ATV as

a complement to their horses, thus keeping a versatile balance between animal and machine if necessary. It's a common sight in rural America to see ATVs towed behind 4 x 4 pickups (the farmer's "big" HMMWV) in a horse trailer with a horses and ATVs loaded up together. If the American farmer and rancher quickly transitioned to the "*Iron Horse*", due to the ATV's versatility, sustainability, agility and cost effectiveness, what happened to the Army's Cavalry branch between World War I and II and why didn't its senior leadership embrace mounted, mechanized warfare as the British, French and especially Guderian did in Germany?

We need to look back to the Army's 'lean years' between World War I and II, when mechanized and armored forces were being developed fully by the European powers and only in fits and starts by the United States. The Army's Cavalry branch faced internal and external challenges – lack of funding, foreign armies developing mechanized forces before the U.S., debate between senior leaders on the definition of "*Cavalry*" and "*Armor*". The older, senior leaders continued to embrace the Webster's Dictionary Definition of Cavalry as Soldiers who fought on *horseback,* versus younger and more far-sighted officers who viewed Cavalry as Soldiers who fought *mounted*[5], which implied the "iron" as well as the "muscle" horse.

The motorcycles of that era and even now, couldn't carry the tremendous amount of equipment and weapons modern ATVs can haul. Two wheel motorcycles and even tricycles, even with sidecars in World War II couldn't have the stability, agility and reliability of the 4/6wheeled ATVs of the nineties have. The closest that the Army ever came to the modern 4 x 4 ATV was the venerable and world famous Ford-Willys Jeep, the _ ton marvel of engineering, who's prototype was built in 49 days by the Bantam Car Company in Pennsylvania In 1940. Over 63,000 were built during World War II. It served in every theater of the war, was floated, hauled, jumped, airdropped, and delivered in gliders. The Jeep was the forerunner of modern ATVs and other "HMMWVs"; the four wheel drive pickup trucks developed and built in the 1950's; which had a tremendous effect on agriculture, industry, hunting and recreational use.

Cavalry and reconnaissance units in the U.S. Army's World

War II grew to depend on the 4 x 4 "ATV" Jeep, which carried 3-4 Soldiers and a pedestal mounted machine gun. The Jeep was truly the first "HMMWV" long before the Hummer would gain fame during the Gulf War of 1991. The reconnaissance platoon of light and medium tank battalions, U.S. Army, had four motorcycles, four Jeeps, one M3 half-track, one officer and 21 enlisted Soldiers.[6] The Army had a mix of platforms, allowing versatility, and is still a reason why motorcycles were considered in the HMMWV MTOE for battalion scout platoons in the nineties. The motorcycle was intended to conduct courier, coordination service and to get into closed terrain the HMMWV's couldn't.[7]

The Jeep was originally designated by the War Department as a "Low silhouette Scout car", certainly true of the famed Jeep and modern ATVs today. That "low silhouette" criteria was set forth intentionally, and stressed stealth and staying low to avoid enemy detection. Unfortunately, both the Bradley CFV and USMC's LAV have very high silhouettes making them easier to detect by the enemy, taller than the HMMWV, itself taller than any ATV and the Ford-Willys Jeep. Private Lloyd Smiley of Great Falls, Montana as a recon jeep driver in the 661st tank Destroyer Battalion, attached to the 69th Infantry Division, part of General George S. Patton's Third Army. Smiley's job was to drive the jeep for his Sergeant, and the jeep also carried a radioman and machine gunner. They would scout as far ahead along roads as possible until making contact with enemy Panzers and then report back to battalion headquarters (the battalion had an entire recon company) on enemy locations so the main fighting force of TRACKED tank destroyers could engage the German armor from an unexpected cross-country axis of advance. The radio was also very effective for calling in friendly artillery support, mortars and P-47 Thunderbolt close air support.[8] He returned to farming after the war, and commented on the benefits of four wheel drive technology for farm use (surplus jeeps were sold and marketed as versatile small utility tractors!) and to this date uses a Kawasaki 4 x 4 1200 pound utility ATV "Mule", even at 75 years of age.

The utility, versatility and cost effectiveness of the modern ATV was quickly picked up by even "I'd never give up my horse" farmers and ranchers. These men and women were brought up through the Depression and World War II, but they were "modern" and young enough at heart to see the tremendous utility and convenience of ATVs. No more would a rider have to bring the horse in, saddle up, carry only limited tools, feed the horse, groom and also worry about his flesh and blood mount getting 'spooked' by a snake or breaking a leg in a gopher hole. No longer would it take years of skilled riding experience to handle a flesh and blood mount, since now only a few weeks and months to establish a minimal ATV riding capability provides more than enough driver skills with risk and safety awareness.[9]

Instead of horse breeds like Quarter Horse, Morgan and Appaloosa, military and civilian ATV riders "ride" Kawasaki, Honda and Polaris. A very common small ATV platform is the Honda Fourtrax 300, weighing 535 pounds, 3.5 gallon fuel tank, four wheel drive, and carries 66 pounds on front cargo rack, double that on rear, not including the rider. A limitation is that most small ATVs use regular gasoline, not diesel, an important logistical criteria, considering the Army's and Department of Defense's intent to go to one fuel, the less volatile JP8.

All United States Air Force and Air National Guard security squadrons now employ small ATVs in their security forces. One ATV and trailer is assigned to each Squad of 13 airmen, and the five-man "flight" (platoon) also has an ATV and trailer. They are VERY sustainable, although still using regular gasoline, but much of the maintenance is performed at the operator level, with operational readiness rates at about 98-100%.[10] The United States Air Force special operations unit has used ATVs, Even the U.S. Army Special Forces (SF) uses small ATVs as Field Manual 31-23 describes ATV employment, battle drills, driver training, navigation, logistics, force packages of 6 x 6 and 4 x 4 ATVs and the best means of delivering and employing ATVs by aircraft for 3-D maneuver. The active 10th SF Group and the Minnesota Air National Guard (active/reserve and joint operation between services) trained together at Camp Ripley, Minnesota and the Air Guard assisted the training in use of wheeled ATVs and tracked snowmobiles, according to Staff Sergeant Joe Hemberger of the Air National Guard.

The British Army, our NATO ally, also has small ATVs (called "quad-bikes" by the British), including the Special Air Service (SAS). The SAS and Long Range Desert Group (LRDG) in WWII used the Ford –Willys Jeep extensively, and were a very reliable recon and attack platform. The U.S. Army Rangers and SAS still operate using Land Rover large ATVs with motorcycles acting as "outriders" and scouts for the main body. Their own "SupaCat" 6 x 6 large ATV, is used by the 5th Airborne Brigade, is a weapon carrier, utility vehicle, transportable by helicopter and is also amphibious, costing about $140,000 in U.S. currency.

ATVs are used in virtually every aspect of American industrial, agricultural, recreational and governmental operations. ATV conventions, jamborees and rallies grow every year, as more and more Americans use ATVs for hunting and recreation. Conduct a web search and you'll see how far the industry has expanded. Consider the following agencies and organizations that use ATVs and ATBs - U.S. Forest Service, Border Patrol, hundreds of state, county and city law enforcement departments, state fish and wildlife agents. Law enforcement in particular prefers the tremendous surveillance, patrolling and reconnaissance capabilities offered by ATVs. ATVs and ATBs also have a positive environmental effect, minimizing man's impact on wilderness areas. That also translates to operational security in a military sense, lower noise signature, tread marks and maneuver damage typical of mechanized operations.

If you consider fuel expenditures alone (remember most small ATVs have ONE cylinder), then that's a great reduction in logistical demand, something the U.S. Army is striving for in its "transformation". Maintenance is ridiculously easy, as many farmers do all the maintenance right in their shops, and just order Worn out parts as needed. The highest level of maintenance would probably be right at battalion, fixed immediately or direct exchanged (ATVs cost $3-7,000, depending on how purchased, as "equipment" by U.S. Air Force, or as a "stand alone" vehicle by itself). Just spark plugs, an occasional oil change, tie rods, brake pads, some wiring – how does that compare to the HMMWV and other larger traditional wheeled and tracked vehicle maintenance requirements? ATV maintenance speaks for itself.

ATVs can also cross water obstacles, and not just be being slingloaded externally or carried internally (A *Blackhawk* helicopter could possibly even carry two ATVs internally, and some design ideas have even suggested "skidloading" to allow better flying conditions). Many ATVs will "float" if accidentally or intentionally flipped over.

The ATV and trailer is designed to help carry the sustainment load, rucksacks and unit equipment of the security airmen, allowing them to be rested and ready for missions and emergencies instead of being "*pack Llamas*" as Army Infantry have long endured. Too often, we see Soldiers LOADED with equipment that could be hauled in their own squads by an ATV and trailer as the Air Force and Air Guard has done. The Air Force pioneered the Carrier Light Auxiliary Weapon (CLAW) system to carry equipment and even emplaced machine guns on ATVs, and also used them as weapons carriers. The Montana Air Guard uses their ATVs for reconnaissance, patrolling, laying wire, medevac and resupply.[11]

The Air Force and Air Guard have operated ATVs successfully in all weather conditions and environments, just as civilian industry has. From 120 degrees plus in Saudi Arabia to the sub-zero windswept airfields of South Korea, the ATV has been very effective and dependable. ATVs were used on every U.S. Air Force deployment (used in Albania, along with Kawasaki Mule, a larger ATV, to traverse massive amounts of mud that bogged down larger supply vehicles) except Somalia, due to no military gasoline available (MOGAS). They were used extensively in joint operations with the Army at the Joint Readiness Training Center, first at Fort Chaffee and then Fort Polk, from 1987 to now. The ATVs are shipped on pallets, and are packaged for immediate deployment globally.

Stealthy Hunting and Scouting

Travel through many of America's forests and game preserves and you will find hundreds of ATVs, trailers and accessories. Hunting uses many of the skills great scouts have – stealth, agility and mobility – and carry enough equipment; weapons and supplies to sustain a "mission" or hunt for several days without re-supply. Hunters STILL go after their prey on foot, and that is still demanded of any Army scout (Military Occupational Specialty 19D, MOS) to be fit enough to scout and fight dismounted. Hunting still requires skill and stealth, but it's certainly an advantage to have an ATV as logistics base, especially if the hunt (or scout/recon mission) continues for several days. The ATV winch is also a tremendous feature, whether pulling a 1200 pound bull Elk out of a treeline or counter-mobility operations typical of scouts, light cavalry and mounted engineers. The modern scout can carry a grenade launchers, automatic weapons, GPS navigation aid, radios, and even a *Javelin* ATGM, whose weight is 49.2 pounds, 35.1 pounds for the missile, and 14.1 pounds for the launch unit.[12] Combine that Anti-tank, Intel gathering and scout abilities with the ability to be moved rapidly and deep by Army helicopters, and you could imagine the tremendous "SHOOT, MOVE, COMMUNICATE and SURVIVE" that an ATV equipped "Light Cavalry" or scout unit could do cross FLOT, or anywhere in the deep, main or rear battle areas.

The Fielding Acquisition of Science and Technology (FAST) team of the U.S. Army's Material Command (AMC) did consider acquiring the John Deere GATOR, a 6 x 6 ATV that had the appearance of a golf cart and utility vehicle to assist Airborne Soldiers in carrying their loads, allowing them to be more rested and ready for combat. The Paratrooper would place his rucksack and additional squad gear on the Gator, and just carry his march and fighting load, not having the be his own "mule" or pack "llama". The John Deere Gator was not as agile and effective as anticipated, but TRULY a step in right direction and innovative. To this date, it appears nothing has happened to find a better system. Must the Soldier continue to 'bear the burden' and remain a "pack animal"? Brigadier General S.L.A. 'Slam' Marshall was very adamant about that issue in his 1950 book, "A Soldier's Load and the Mobility of a Nation", and it seems we've learned little. We have developed "ultra-light" fabrics, equipment, tools (recreational users have "state of art" equipment, very costly, but certainly the American fighting Soldier deserves), but commanders and staffs forget that '100 pounds of ultra-light equipment' is *STILL 100 pounds of equipment.'*

The Combat Support Vehicle (CSV)

The ancient Roman Army at its smallest 8-man unit level, had a mule to help transport the Legionnaire's equipment and tentage, all today's U.S. Army infantry Soldier has his own back. The Army at this very moment could purchase the same package as U.S. Air Force has for it's "infantry", and finally relieve the burden of the infantry Soldier by putting his existence load on a trailer right in his own squad. This act would NOT Make him any weaker, but preserve his strength and vigor to WIN when he reaches the attack or battle position, ready to assault with energy or prepare the battle position. For less than the COST of an anti-tank missile, finally the U.S. Army Light Infantry Squad (Airborne, Air Assault, Light and Rangers) has a Combat Support Vehicle (CSV) that can go anywhere they can go that's cost effective, easy to maintain, uses very small amount of fuel, and can be AIR DROPPED and carried by every U.S. Army Aviation platform.

Polaris Industries of Minnesota has developed the first Diesel equipped ATV, the Sportsman 500. It's 3.75 gallon fuel tank offers about 20 miles per gallon (dependent on terrain), 50.5 inch wheelbase, dry weight of 755 pounds, height of 47 inches, width of 46 inches and length of 77 inches. Many ATVs also have many accessories such as winches and trailers. The winch for the Sportsman 500 has a 2,000-pound capability, a very useful tactical and utility feature, especially for ATV mounted combat engineers. The engine is .455 liter and water-cooled. The rear cargo rack load capacity is 180 pounds and front rack is 90 pounds.[13] The Sportsman costs about $7,000 retail and can also tow 1,225 pounds. Many gasoline powered ATVs also have similar utility and performance features, but to this date, Polaris is the only one with a diesel. Fuel gelling and hard starting in cold conditions can be overcome with anti-gelling agents and Glow plug or cylinder heaters. The Diesel ATV has greater torque, but lower speed, an important factor, but its diesel fuel capability should be heavily favored. Improvements to increase speed are certainly feasible, given enough time and interest.

This innovation is a vital component of *Air*-Mech-Strike

Doctrine to enhance light infantry mobility. Mounted infantry in *Gavin* MAVs and *Ridgways* have their sustainment loads carried on the APC and AFV, no significant change since Mechanized infantry was fielded during the 1930's. Not only does the "CSV" enhance mobility, but is also a great combat multiplier. It could be used just as Air Force and Air Guard security units employ them now – but also for reconnaissance, communications, laying landline, emplacing obstacles, delivering supplies and medevac. This also has great utility in Support and Stability Operations (SASO), where the smaller ATV can negotiate smaller, foreign streets, villages and wilderness trails, whether East Timor or Kosovo.

The lack of armor on an ATV is unavoidable and certainly a significant concern, but allows the "mounted" Soldier to take advantage of the new Interceptor Body Armor and wear the complete ensemble, since his weight is being borne by the ATV itself. Soldiers have to receive adequate training on ATV handling and safety and must train and conduct battle drills for reacting to ambushes and immediate dismounting, like his cavalry forebear. Effective reconnaissance, situational awareness and stealth all are the first key component of protection – avoiding detection and finding enemy first is critical to the Army's Initial Brigade Combat Team (BCT) concept, using high technology, recon platoons and an entire Reconnaissance, Surveillance and Target Acquisition Squadron (RSTA) to find the enemy and develop the situation.

FLYER 21 – The 21st Century 'Jeep'?

Enter the *FLYER 21*, a cross between the M274 "*Mule*" and the "*Hummer*", with the potential to become the "Ford-Willy's Jeep" of Force XXI – lighter, deployable, sustainable, agile and highly DEPLOYABLE. The *FLYER 21* utility vehicle, produced by the Flyer Group Inc., Marina Del Rey, California, has been developed, tested, is in production and in currently in military use outside the United States. The *Flyer 21* family of vehicles is in no way intended to replace the highly successful HMMWV series, but offers to complement it. The *Hummer* is a true "workhorse", extremely useful for logistics, heavy weapons platforms, medevac, maintenance, Command and control and other utility tasks. With over 70,000 built and most still in use, the *Hummer* will still be a vital asset to the Department of Defense.

Although the HMMWV has been proven more capable of reconnaissance than the *Bradley* CFV, it still wasn't specifically designed to replace the _ Ton scout vehicle, the M151A2 jeep. Consider basic physics – how do you expect a **1 _ Ton** Truck, the Hummer, to meet the same stealth and "low silhouette scout car" (War Department's title specification for the "Jeep") characteristics that a **_ Ton** vehicle? The *Hummer* has tremendous agility *for it's size,* but is it still quieter, lower, smaller and lighter than the initial Ford-Willys Jeep and slightly less capable descendent, the M151 jeep series? The Israeli Defense Force (IDF), proved that their Anti-Tank jeeps were very effective during the 1973 "*Yom Kippur*" War, when the IDF needed every fighting platform it could muster to counter a surprise attack. Why didn't we just design a smaller "HMMWV" initially, or was everything focused on "heavy" *Fulda Gap* force structure – since the current HMMWV *looks "small"* as the famous jeep when compared next to the huge M1 *Abrams* Main Battle Tank and *Bradley* Fighting Vehicle (note; fighting, not reconnaissance). One "size" doesn't necessarily fit all functions and tasks, multi-purpose or not, no matter how inventive.

The *Flyer 21* itself was conceived and built upon the successful design concepts and technology incorporated into the Hummer itself. It is significantly lighter than the HMMWV, with a curb weight of 3,000 pounds and a payload of *3,000 pounds.* The Turbo-charged intercooled 110 HP diesel engine is mounted in the rear. It has four wheel disc brakes and independent suspension, uses tubular space frame, and has 36 x 9 x R16 *Michelin* or *Goodyear* tires. The *Flyer 21* is also an "open" vehicle and has no roof or side curtains, although American" 'field ingenuity" could provide that feature in assembly and staging areas.

It can carry 4-6 Soldiers depending on seating and mission configuration, and can support many weapons - .50 cal M2 Machine Gun, TOW, *Javelin* Anti-tank Guided Missiles (ATGMs), MK19 40mm Grenade launchers and other combat support equipment, technology and digitalization. It's performance characteristics include: 20 gallon fuel tank (optional 70 gallon, *allows 1,000 mile* range) that provides 325 mile cruising range; can negotiate 60% grade, 50% slide slope, 16 inch vertical step and ford 24 inches of water with no preparation. Maximum speed is 65 mph, a valuable factor to get clear of indirect and direct fire. The *Flyer 21*'s length is 179 inches, width 63 inches and overall height 72.5 inches (61 inches when rollbars folded). It has 18 inch ground clearance and it's tires have the optional capability of "run flat" technology. The Flyer series family of vehicles meets the Joint Operational Requirement Directive (JORD) and meets military specifications, particularly for Light Strike Vehicle (LSV) requirements. That INCLUDES being transportable by CH-47D/F and CH-53 helicopters.[14] The combat weight of the Flyer fully loaded at 5,500 pounds can also be lifted by the UH-60 *Blackhawk* helicopter, a vital part of *Air*-Mech-Strike's effectiveness one of it's very unique features is its ability to stack one atop another, allowing **SIX Flyer vehicles to be stacked into a C130 aircraft**. The CH-53 and CH-47D/F *Chinook* can both carry two internally and externally, and the UH-60 Alpha or Lima models can carry one or two, either combat loaded or curb weight.

Principles of *Air*-Mech-Strike Reconnaissance

Before going further and into scout platoon organizations, the

Flyer 21 ATVs (Flyer Group and Carol Murphy)

tasks and purpose of Reconnaissance must be thorough examined and stressed. Too often, even experienced operators confuse screen, guard, security and reconnaissance missions, leading to misunderstanding and false expectations of cavalry and battalion scout platoons.

The following reconnaissance principles are:
Maximum Recon Forward
Orient on Recon Objective (named areas of interest/avenues of approach)
Rapid and Accurate Reporting
Retain Freedom to Maneuver
Gain/Maintain Enemy Contact
Develop Situation Rapidly

The Following missions are:
Route recon   Area recon   Zone recon
Screen main body   Passage of Lines

    Scout platoon capabilities include conducting liaison, quartering parties, traffic control, NBC survey, pioneer and demolition work and participate in area security.[15] FM 17-98 does make distinctions between which platform does certain tasks and missions best between the wheeled HMMWVs and the tracked *Bradley* M3 Cavalry FIGHTING vehicle (CFV). The *Bradley* CFV is not deployed in battalion scout platoons, but in cavalry troops with tank platoons. As stated already, the wheeled recon platforms were considered superior for strictly reconnaissance missions, but retaining the freedom to maneuver was certainly a Bradley M3 CFV strength considering it's firepower.

    Another consideration for scout platoons is SASO missions. The unique flexibility of Flyer/ATV equipped scout platoons allows the platoon to negotiate more narrow streets and trails in those type operations, more flexibility. The lack of armor could be a liability, depending on the perceptions of the local population, history, etc.

Air-Mech-Strike Battalion Scout Platoon MTOE Options

The MTOE options suggested here are viable for current force structure or the *Air*-Mech-Strike concept, and could be done immediately. The scout platoon options also work for the Army's Initial BCT scout platoons and offer equal if not greater capability at lower cost—with $880,000 budgeted for just one "common platform" "Medium Armored Vehicle" (MAV) we could outfit an entire Flyer/ATV hybrid scout platoon with all its equipment, weapons, and UAV. The suggested MTOE was designed to keep leader and scout disruption to a minimum and also use current weapons and equipment in the Army inventory. The HMMWV that would be drawn out of platoons could easily be redirected to many other units and duties.

The scout platoon of wheeled Flyers and ATVs would operate with the Wiesel 2 if they are field in the "light cavalry" RSTA squadron, a parallel cavalry organization where *Air*-Mech-Strike replaces tanks for Ridgways as the armor platform and utilize the Flyer/ATV scout platoon in lieu of the Bradley M3 CFV of the heavy cavalry troop. Again, the light cavalry should NOT conduct guard missions, like an armored (heavy) cavalry Regiment mission.

The following options are structured around the infantry battalion scout platoon as a "base" element and thoroughly integrates ATVs into MTOE, as originally intended for motorcycles *(ATV should be adapted immediately, regardless of current force transformation to fulfill role in MTOE for motorcycles).* Hummers tapped to be replaced will go elsewhere in the force structure, particularly for command/control, heavy weapons and logistical platforms.

Screening criteria was built around minimizing disruption to current MTOE, Soldier MOS, weapons and equipment all standards to current U.S. Army force structure, from MK19 employment, LP/Ops and slingloading techniques. The long awaited Long Range Acquisition System (LRAS, with it's CCD TVs, FLIR and other high technology) is certainly part of all options under C4ISR, but not identified in great detail, but certainly required. ALL scout platoon Soldiers should wear IN-

TERCEPTOR rifle-caliber resistant body armor, complete ensemble, the standard for the Army's Land Warrior program and all MUST have Soldier Intercom system. Special camouflage modules for all platforms are assumed, to include anti-ballistic blankets, to improve force protection at every opportunity. An anti-armor role is possible, similar to previous Tow Light Anti-Tank, "TLAT" concepts that were common in ARNG, and the German Airborne Anti-tank Battalion with *Wiesel* AFV[16] using *Javelin* fire/forget ATGMs without any vehicle modification.

Flyer/ATV Composite Squads, BN Scout PLT HQ Minus ATVS

1. Replace 10 Hummers with Six Flyer 21 "large" ATV (can be delivered in six months off production line)
    2. One Flyer per squad (1 x 4), five scouts per squad, add two Polaris Diesel ATVs and scout riders for total of seven 19D scouts in squad. Each Flyer has collapsible trailer for towing TWO small ATVs or mounting brackets to carry two folded ATBs.
    3. Every Flyer carries *Javelin* ATGM and three rounds minimum, GPS, SINCGARS and "state of the art" C4ISR equipment and lightweight anti-ballistic shielding.
    4. Each section (2 squads x 2) has MK19 40mm grenade and .50 cal M2 Heavy Machine Guns.
    5. Platoon HQ same without ATVs and riders, still has UAV, UAV operator and medic. Trailers still in HQ, used for LOGPAC and moving four ATVs on each, One combat lifesaver per Flyer.
    6. Platoon can carry two extra passengers (Forward observers, COLT team, etc, dependent on mission and situation); Platoon Sergeant has to coordinate to get any additional vehicle to transport ATV riders from the squads if platoon engaged in long motor march. Platoon strength is 28 in four squads, plus five in Plt HQ, and medic = 34, greater than current Hummer Scout MTOE.
    7. Six Flyers (4 on 4 stacking), eight ATVs, six collapsible trailers, UAV (approximate weight, combat loaded platforms and troops, equipment = 41,400 pounds/2 = 20,700 pound load for one C130 sortie allowing maximum aircraft cruising range) and unit basic load (UBL) can deploy in TWO sorties (Hummer

platoon would take at least FOUR sorties); Class III fuel to refuel all platforms approximately 200 gal, 1,400 pounds.

8. One *Blackhawk* helicopter company could lift entire platoon in one serial, without significant operational range degradation.

9. One squad – *Flyer 21* team (squad leader, driver and gunner), two dismounts (or one, but always work as two, take driver from Flyer) and always TWO ATV Riders, offering capability to cover THREE recon objectives, for a platoon total of (4 x 3) of 12 recon objectives, still greater than current *Hummer* benchmark of eight.

10. Cost: 6 Flyers x $100K = $600K (still cheaper than Initial BCT budget of $880K for just one MAV recon platform with five scouts, six per BN scout platoon in Initial BCT); eight Polaris Diesel ATV x $7K = $56K: add UAV, C4ISR enhancements, four more trailers, lighter anti-ballistic shielding ($100K?) for conservative total of $758,000 dollars, still under MAV BCT budget for one platform, but needs FOUR more Soldiers.

ADVANTAGES: Option 50% (12) more recon objectives than HMMWV MTOE; deploys in two C-130 sorties complete; entire platoon can be airlifted by a *Blackhawk* helicopter company without operational degradation, greater mobility at squad level; Plt HQ can add combat support teams like COLT

DISADVANTAGES: greater cost than current *Hummer* MTOE, but still cheaper than one BCT MAV 'recon' platform; may require additional vehicle to move ATV riders on long motor march in theater.

HUMMER/ATV: Interim Solution, Particularly for Reserve Components

1. Retain Six *Hummers* (M1025/1026; send *XM1114 back* into force as armored up command and control platform in heavy armor fighting element, reduces logistical burden)

2. One *Hummer* per squad, five 19D scouts per squad, add TWO Polaris Diesel ATV riders with collapsible trailer, total of seven scouts in squad.

3. Every HMMWV carries *Javelin* ATGM and three rounds minimum, GPS, SINCGARS, and 'state of art' C4ISR equip-

ment and lightweight anti-ballistic shielding, **not** the full up-armored XM1114 package.

4. Every section (two squads x 2) has MK19 40mm grenade launcher and .50 cal M2 Machine gun. Platoon Headquarters same as option two, but uses *Hummer* instead of *Flyer 21*, has supply trailers UAV and operator, medic and can carry combat support team like COLT. Four squads x 7 scouts = 28, plus six in Plt HQ (not counting COLT), for grand sum of 34 scouts.

5. Six HMMWV M1025/2026 platforms, eight ATVs, could be deployed in THREE C-130 sorties with all equipment, troops and unit basic load, maybe two dependent on METT-T. Heavier weight of *Hummer*, even if NOT armored XM1114, could be lifted by UH-60 Lima model. Platoon could be lifted by CH-47D helicopters by 4-6 helicopters in one serial, METT-T dependent.

6. One squad, mounted (3), dismount team (2) and ATV team (2) could cover thee recon objectives, a total of 12 for the platoon, greater than current *Hummer* MTOE.

COST: Significantly cheaper, eight ATVs at $7K = $56K, plus additional technology, C4ISR, armor, and equipment = $80K, a total of $136K, about 15% the cost of one BCT MAV "recon" at $880,000 per.

These options, and the advantages/disadvantages, do relate to the Army's Transformation Force Characteristics – particularly in mission accomplishment, lethality, agility and deployability. Lethality is achieved with the ability to carry the new *Javelin* ATGM, greater precision targeting of joint fires via digital commo, COLT teams, attack helicopter (*Air*-Mech-Strike OH-58s if necessary, and Apaches from the Corps Commander) and USAF close air support. Survivability is achieved by being agile and mobile, by both air and land and avoiding the enemy's strength through robust Intel and recon. Sustainability of ATV is particular is a tremendous advantage, they have passed the very demanding requirements of civil, industrial and agricultural use. Their versatility makes them truly "highly mobile, MULTI-PURPOSE wheeled vehicles" and they have unmatched mobility for scouting, hunting and search and rescue. The simple combat maxim criteria of 'shoot (observe and report for recon, can still be 'lethal to break contact) move (by land and air), communicate (leadership) and SURVIVE (protection) are all met by the *Air*-Mech-Strike BN scout platoon options in varying degrees.

The value of wheeled reconnaissance platforms has been proven, based upon the tremendous success and versatility of the original "low silhouette scout car"[17] during World War II, Korea and Vietnam, even as patrol jeeps by units of the 1st Infantry Division. The modern ATV has come of age in the nineties, truly a "workhorse" that can be increasingly exploited for military use as the U.S. Air Force, Air National Guard, Special Operations Command (SOCOM), British Army and the German Bundeswehr have.

The U.S. Border Patrol is using ATVs and ATBs to patrol our Nation's borders and conducting surveillance with them. Even the Army of Singapore possesses Flyer 21 'ATV/HMMWVs and employ them with their helicopters in their own *Air* -Mech-Strike application. Non-digitized *Air*-Mech-Strike! is a reality - the British 5th Airborne Brigade did it when moving into Kosovo in July 1999, lifting it's *Scimitar* tanks and SupaCat 6 x 6 ATVs with Royal Air Force CH-47D *Chinook* helicopters, avoiding thousands of mines, obstacles and unnecessary casualties.[18]

---

[1] U.S. Army Field Manual (FM), 100-5, *Operations*, (Washington D.C: U.S.Government Printing Office June 1993). pp. 2-10, 2-11

[2] Ibid.

[3] Bob Sillary, *The Plane that Drove*, *Popular Science*, March 2000, pp. 72-74.

[4] Lieutenant Colonel Liebert's personal experience, operator, Windwalker Ranch, Great Falls, MT and BS, Agriculture, Purdue University regarding agricultural ATV applications and personal use.

[5] Robert W. Grow, Major General (Retired, U.S. Army), *The Lean Years*, (Fort Knox, Kentucky: U.S. Army *Armor* Magazine, Jan-Feb 1987) p. 23.

[6] U.S. Army Field Manual (FM), 17-33, *The Armored Battalion*, Sep 18, 1942, 15.

[7] 1LT Charles W. Gameros, Jr., *Scout HMMWVs and Bradley CFVs*, (Fort Knox, Kentucky: *Armor* Magazine, Sep-Oct 1991) p. 22.

[8] Interview with Lloyd Smiley, *Veteran Recounts WWII*, (Great Falls, Montana: *Great Falls Tribune*, 4 April 1999) p. 5.

[9] SSG Joe Hemberger, ATV Safety Trainer, Minnesota Air National Guard, e-mail to Lieutenant Colonel Liebert, 24 February 2000.

[10] Ibid.

[11] Master Sergeant Eric. Holman, ATV expert, email to Lieutenant Colonel Liebert; Lackland AFB, 30 Aug 1999, *eric.holman@lackland.af.mil*

[12] MAJ Bradley N. McDonald, Javelin, (Fort Benning, Georgia: U.S. Army *Infantry* Magazine, September-December 1998) p. 5.

[13] Interview with Robert Anderson, Special Markets Director, Polaris Industries, 28 Nov 1999.

[14] Interview with Bob Parker, Flyer Group Inc, 17 Dec 1999 and Flyer Group literature, Oct 1999; *flyergroup@aol.com*

[15] Field Manual 17-98 (FM), *The Scout Platoon*, (Washington D.C.: 1994 U.S. Government Printing Office).

[16] LTC Wolfgang Mettler, German Army, *The German Airborne Anti-tank Battalion*, (Fort Benning, Georgia: U.S. Army *Infantry* Magazine, January-February 1995). pp. 24-28.

[17] U.S. War Department "design specification", 1940.

[18] Major Chuck A. Jarnot, USA, *Air-Mech-Strike*, (Arlington, Virginia: *Army* Magazine, Jan 2000.

Air-Mech-Strike 3-D Maneuver Armored Fighting Vehicle Study

Capability	Rubber-Track AFVs	Metal-Track AFVs	Rubber-Tire AFVs
	4-ton RFVs	11-ton M113A3s	16.3 ton LAV-III armored
	10-ton GFVs	17-ton M8 AGSs	8x8 car
C-747F transportability	YES RFVs/GFVs(?)	Maybe	NO has High Axle weights
UH-60L transportability	Yes	NO	NO
CH-47F transportability	Yes	NO	NO
C-130 transportability	Yes	Yes	NO large turrets
C-17 transportability	GFV: 5, RFV: 10-12	GFV: 5, MTVL/AGS: 4	3
Parachute airdrop	Yes	Yes	NO
Route flexibility	Yes	Yes	NO
Ground pressure	RFV: 4 PSI	M113A3: 8.63 PSI	4 tons per axle
X-country mobility/SWIM	Yes/Yes	Yes/Yes	NO/NO-GO go up steep banks
Gap/obstacle x-ing	RFV: 48" gaps, 24" vert walls	MTVL: 86" gaps; 36" vert walls	66" gaps, 24 inch vert obstacle
	GFV: 66" gaps, 36" walls	M8 AGS: 86" gaps, 36" vert walls	
Power-to-weight ratio	RFV: 31hp/ton	M113A3: 20.2 hp/ton	20 hp/ton
	GFV: 22 hp/ton	MTVL: 22 hp/ton, AGS: 28hp/ton	
Fuel economy	RFV: 7-13 mpg	M113A3: 3.5 mpg	3.6 mpg
	GFV: 4.0 mpg	MTVL: 3.0 mpg	
Need truck transporters?	NO	Maybe	NO
Maneuver/turn radius	0 meters Pivot turn	0 meters Pivot turn	17 meters U- turn
Road speed	55 mph	41 mph	62 mph
Survive Enemy small arms fire	No damage	No damage	run flats: 5 miles @5mph
O & M costs	----------same as wheels----------		
GVW,volume, payload growth?	YES---------28% more efficient than wheels		NO
Acquisition costs	GFV: $150K	M113A3: $0-150K	LAV: $880K
	RFV: $400K	M8 AGS: $3.5 MIL	
U.S. Army service	Yes, RFV tested	Yes, AGS classified as M8	NO

Chapter 8

3-D Air-Mech-Strike Organization

*"Air-Mobile units will not fight battles alone,
but as an integral part of the land forces,
along with armored and other ground forces"*
General John R. Galvin (Retired) in *Air Assault:
The development of Airmobile Warfare*

A Brigade Combat Team (BCT) capable of 2-D/3-D combined-arms firepower and maneuver would best be configured by creating a *Gavin* Infantry Battalion capable of 3-D maneuver, keeping one M1 Armor and one M2 BFV Infantry Battalion capable of 2-D maneuver, supported by a *Ridgway* (*Wiesel*) Reconnaissance Squadron capable of optimal 2-D or 3-D movement to gain information to plan/execute operations. Instead of "starting all over" and creating entire Interim Brigades with a single newly purchased vehicle type, the *Air*-Mech-Strike model integrates the best aspects of heavy, light and "medium" 2-D/3-D capabilities in every BCT providing optimum flexibility. Former NATO Commander General John R. Galvin (Retired) in his important work, *Air Assault: The development of Airmobile Warfare* writes: *"The plains of western Europe, however, are not as flat as the Germans made them seem with the blitz against the Low Countries...The plain is full of obstacles, especially water barriers, that can be used by Air-mobile force to assist either offensive or defensive tactics. It is easy to forget how important were the bridges of the Seine, the Meuse and the Rhine. An Army Corps containing an Armored Division, a Mechanized Infantry Division, and an Air-mobile Division would provide unparalleled power and versatility in combat anywhere in Europe."*

In Air-Mech-Strike, we take these principles and build powerful 2-D/3-D capable combined-arms Divisions, starting with their Brigade Combat Teams maximizing existing equipment, leveraging off-the-shelf technology with modest purchases and exploiting the U.S. Army's superiority in helicopter numbers and types.

Brigade Combat Teams One Battalion at a Time

In lieu of building just a few expensive BCTs "from-scratch" over a long period of time, we recommend the entire Army be reorganized for flexibility, changing selected Battalions to the *Air*-Mech-Strike model, and keeping the other Battalions "as is". This way, the entire Army will be available for full-spectrum operations, reducing operational tempo and strain on personnel[1]. For the cost of five BCTs we could have the entire Army in full-spectrum, 2-D/3-D *Air*-Mech-Strike capable BCTs.

The M1 *Abrams* tank heavy Battalions dominate the 2-D close fight, and provide a breakthrough capability and maneuver[2]. The BFV Battalion in the BCT provides close-in security and ground-taking action along the 2-D axis of advance. A full 6 x M1064A3 120mm *Gavin* heavy mortar platoon would provide organic indirect fire support for the BFV and Tank Battalions[3].

Since the M1/M2 heavy, large AFVs are resource-intensive when deploying by USAF aircraft and cannot deploy by Army aircraft, the *Gavin* infantry Battalion and the RSTA Squadron provide flexibility to every Brigade and Division since they can deploy in large numbers by aircraft and conduct 3-D maneuver on the battlefield[4].

Small *Ridgway* AMVs based on the German *Wiesel* were selected for the BCT RSTA Squadrons to enhance reconnaissance and surveillance. The small RFV also provides emergency direct fire support for dismounted recon.

The utilization of existing *Gavins* and conversion of other M113s to *Gavin* standard will save the Army millions of dollars and these savings can "pay" for the purchase of small *Ridgway* AMVs in the RSTA Squadrons.

For the cost of five BCTs "from scratch", we could have all 10 Divisions of the active Army Component, and begin to convert a substantial portion of the Army National Guard and Army Reserve to the *Air*-Mech-Strike model. Only selected Battalions are converted, not the entire Brigade. The transformation of the Army 's Divisions could be accomplished within 4 years.

Possibilities include having existing BFV turrets fitted to selected stretched *Gavins* to create *Gavin* Fighting Vehicles (GFVs). All GFV infantry carriers have ACAV gunshields, side-mounted Medium Machine Guns, and a Heavy Machine Gun or a Mk-19 Grenade Machine Gun until a 30mm M230 autocannon/*Javelin* ATGM cupola is fielded.[5] Elevated TOW ATGM, 120mm gun-mortar, 25-105mm gun turrets fitted to regular (Infantry Tactical Vehicle Light-ITVL) or stretched-hull variants (Mobile Tactical Vehicle Light-MTVLs) become *Gavin* Weapons company systems to provide short-range direct fire support. To date, the best direct-fire support platform is the M8 Armored Gun System which has a shoot-on-the-move 105mm gun capability to win meeting engagements with enemy armor—the AMS BCT force structure has the means that the 9th Infantry Division lacked in the 1980s. The M8 AGS is type-classified; has a large stockpile of 105mm rounds to include anti-personnel "*beehive*", white phosphorous; Armor-Piercing Discarding Sabot, kills all known enemy tanks and high-explosive to blast buildings, bunkers and create breach holes for infantry to assault buildings. The 105mm round has a 35% growth potential to keep it a viable tank killer for many years into the future. When developed/fielded, 105mm - 155mm howitzers or HIMARS 227mm rocket launchers on Palletized Loading System flat racks, can be fitted to XM1108 *Gavin* variants, which have cut-down bodies and become self-propelled long-range artillery means for the BCTs[6].

The RSTA Squadron would be outfitted with the off-the-shelf small *Ridgway* AMV based on the German *Wiesel* in the 3-4 ton weight range, and would be a new purchase vehicle for the Army.

One Infantry or Armor Battalion in each Brigade of every Army Division converts to *Gavins or Ridgways*.

Every Brigade in every Division reorganizes its Scouts or Cavalry troops from armored HMMWVs to small *Ridgway* AMVs based on the German *Wiesel* in a RSTA Squadron.

Excess M1/M2 heavy AFVs and trained 11M, 19-series personnel freed up in the conversion to *Gavins* transfer to the RSTA Squadron or the 4 Light Infantry Divisions to be RFV and GFV crewmembers. Excess M1/M2s beyond this transfer to the National Guard to speed up conversion to the heavy model if designed. Other BFVs are exchanged for selected M113A3s in heavy engineer units.

New Equipment Training (NET) on *Gavins* will be conducted for selected Battalions in the Army's light Divisions, and new 11M/19-series personnel in *Gavin* Battalions will be stood up in Heavy Divisions to fully fill the ranks with trained, combat-ready Soldiers.

Low-cost off-the-shelf diesel ATVs and trailers are supplied as "Combat Support Vehicles" to every Squad in the Army's Light Division's; the 2 Battalions per Brigade not equipped with *Gavins* to solve the Soldier's load problem. Other ATVs/ATBs supplied to RSTA Squadrons to improve stealthy mobility with the RFV AMV as a "mother" vehicle.

80% of Army Aviation Consolidated in Corps Aviation Command

Most of Army Aviation would be consolidated into COAVCOM and assigned one each to a warfighting Corps. The one exception is a single squadron of UH-60s and a single *Kiowa Warrior* troop. *Air*-Mech-Strike requires timely concentration of Army Aviation assets. The range that typical *Air*-Mech-Strike operations occur will require the numbers of airframes available to be consolidated into a Corps organization. The COAVCOM would be sub-organized into multi-functional Aviation Brigades with functional regiments of lift, attack and maintenance. Deploying helo packages would be developed in echelon at the COAVCOM. The ability to tailor the aviation packages needed from Corps level would greatly enhance the *supported unit's* flexibility. Army Airspace Command and Control (A2-C2) would be enhanced by elevating the airspace command decision process to a Divisional level staff that most likely owns most of the helicopters that would be operating in the Corps area of operation.

Divisional Aviation Squadron

The one Divisional helicopter unit would be organized as an

Aviation Squadron with 18-24 x UH-60L helicopters, 8 x OH-58D *Kiowa Warriors* armed reconnaissance helicopters and one or more RFV and ATV ground troops. The Aviation Squadron at the Division would provide the UH-60 lift to conduct insertions of RSTA units and Air Assaults of infantry, ADA, and other forces. The organization should have at least a second Reserve Component crew per aircraft enabling them to generate more sorties per 24-hour period. The UH-60s should come with *Hellfire* and *Hydra-70* rocket pods for their wing stores (ESSS). This capability is well developed and utilized in Army SOF on their AH-60L Direct Action Penetrators[7]. The OH-58D can identify targets and "laze" (laser target designate) for the UH-60 *Hellfires* (16 X missiles per aircraft with a full reload inside, or 76 *Hydra 70* rockets), AH-64s, guided 155mm artillery projectiles and USAF TACAIR ordnance. On other missions, the UH-60s can transport troops and/or tracked RFV AMVs or wheeled ATVs while the OH-58Ds can target or fire their own *Hellfire* missiles and *Hydra-70* rockets[8].

The Army National Guard and the Army Reserve

From the end of the Cold War to present, the *Citizen-Soldiers* that make up our Nation's Reserve Component (RC) has picked up more and more of the land warfare equation. Their challenge with adapting to the complex rigors of implementing *Air*-Mech-Strike 3-D maneuver is their limited time during drills and annual training. With only about 40 days a year to train the RC should approach *Air*-Mech-Strike in a deliberate long-term process. Until a future combat vehicle and future aircraft can be built that supersedes all of our present systems, heavy armor is an essential ingredient to the Army's force structure. The length of time it takes to move by sea a heavy Brigade or Division to a Major Theater War in South West Asia or Korea provides a "*training window*" to exploit for RC units. This suggests that RC combat units make up the majority of the Army's heavy forces. The M1 and M2/3 systems that *Air*-Mech would displace could accelerate the fielding units currently scheduled for conversion. The freed up M113s could provide depots with a stockpile of vehicles for conversion to *Gavin* Fighting Vehicles (GFVs).

Where affordable, Army National Guard and the Army Reserve combat units should mirror their Active Component (AC) counterparts. Enhanced Brigades in particular, would be the logical choice for the implementation of AMS.[9] Much of the required Army Aviation units required to effect AMS are found in the RC.

Division Re-Organization

The Army's Divisions both active and reserve would reorganize to one of four models as depicted in the enclosed charts:

(AMS) *Air*-Mech-Strike Symbology
Army Combat Division: Heavy conversion to AMS
Army Combat Division: Light conversion to AMS
Army Combat Division: Airborne conversion to AMS
Army Combat Division: Air Assault conversion to AMS

Note that different arrangements in the Brigade Combat Team are used for the *Gavin*, *Ridgway* and CSV enhanced Battalions to optimize warfighting capabilities. In the Heavy Combat Divisions, we add a lighter weight armored vehicle in each combat brigade and the RSTA Squadron to provide 3-D maneuver and reconnaissance capabilities. In the Light Combat Divisions, we add a lightweight armored vehicle in each combat brigade and the RSTA where there was no armored capability. CSVs are supplied to enhance infantry mobility in the other infantry Battalions. The Airborne Division is the same structure as Light Divisions except all parts are capable of parachute delivery. The Air Assault Division is the same except that the *Ridgway* is used as the base vehicle for the AMS capable infantry Battalions in order to be UH-60 helicopter transportable.

Gavin Fighting Vehicle Infantry Battalion

3 x Rifle Companies of 14 x *Gavin*-MAVs, plus BN CO/XO x G-MAVs = 44 x *Gavins*. The *Gavin* Battalion would be closely organized to mirror the numbers of its brother M2 BFV Battalion so infantry leaders can be assigned to either outfit with minimal train-up.

Gavin Headquarters and Headquarters Company

The main differences between a M2 BFV Headquarters and Headquarters Company (HHC) Modified Tables of Organization and Equipment (MTOE) and a M113A3 Mechanized Infantry Battalion HHC MTOE would be their *Gavin* 120mm mortar squads are in the *Gavin* Weapon Company (Has a Mobile Gun System or Armored Gun System, and Anti-Tank *Gavins*), a Sniper Detachment assigned to the Battalion Scout platoon in HMMWV 4x4 trucks, and less maintenance personnel/vehicles. The Sniper Detachment would have 6 x 2-man Sniper Teams with a Sniper Detachment officer and detachment Sergeant as the C2 element.

Gavin Rifle Company

3 Platoons of 4 x GFVs, plus Co/XO x GFVs per Company = 14 x GFVs. Since a Gavin only needs 2 crewmen, BFV units converting to *Gavins* have their BFV Commanders become *Gavin* Track Commanders. BFV Drivers become *Gavin* Drivers. BFV Gunners could become *Gavin*-MGS or AT crewmen or *Gavin* infantrymen.
    2 Crewmen, 9-11 dismounting infantry seats per GFV
    4 GFVs x 9-11 dismounting seats = 36-44 seats available
    4 x 9-man Squads of Dismounting infantry per Platoon = 36 seats required
    +1 Platoon leader = only 37 seats required
    +RTO = 38 seats required
    +Medic = 39 seats required

Gavin Rifle Platoon

A Gavin-MAV with gunshields for the *Gavin* Commander and 2 Machine Gunners is a "*Gavin* Fighting Vehicle" or GFV.
    <u>*Gavin* Fighting Vehicle 1</u>
    Gavin Commander mans .50 caliber HMG and MK19 40mm Grenade Machine Gun from cupola
    Driver
    1. 1st Squad Leader

2. Team Leader A

3. AR mans M249 LMG with quick-release mounts from top troop hatch gunshield

4. Grenadier

5. Rifleman

6. MMG Gunner B

7. MMG AG B

8. AT Team A Gunner *Javelin* "fire/forget" ATGM or 84mm M3 RAAWS "*Carl Gustav*" on pedestal mount

9. Assistant AT Soldier loads *Javelin* or 84mm rounds for Gunner

10. Platoon leader with CVC helmet commands from top troop hatch, dismounts and fights the Platoon with the Weapons squad as his foot-mobile base-of-fire able to accompany them where vehicles may not be able to follow

11. RTO

Gavin Fighting Vehicle 2

Gavin Commander mans .50 caliber HMG and MK19 40mm Grenade Machine Gun from cupola

Driver

1. 2d Squad Leader

2. Team Leader A

3. AR mans M249 LMG with quick-release mounts from top troop hatch gunshield

4. Grenadier

5. Rifleman

6. Team Leader B

7. AR mans M249 LMG with quick-release mounts from top troop hatch gunshield

8. Grenadier

9. Rifleman/Sniper with magazine-fed 7.62mm gas-operated rifle

Gavin Fighting Vehicle 3

Gavin Commander (Platoon Sergeant) mans 25mm autocannon/7.62mm MMG turret leading the other *Gavins* as the mobile base-of-fire element.

A one-man 25mm Bushmaster/7.62mm co-axial Medium

193

Machine Gun turret could be fitted in the location of the TC and a full 9-man dismounting infantry squad carried—in fact there is room for one more Soldier—for a total of 10. An integrated helmet sight system like used on attack helicopters allows the *Gavin* Commander to direct the driver, land navigate and engage targets with his gun turret.

Driver
1. 3d Squad Leader
2. Team Leader A
3. AR mans M249 LMG with quick-release mounts from top troop hatch gunshield
4. Grenadier
5. Rifleman
6. Team Leader B
7. AR mans M249 LMG with quick-release mounts from top troop hatch gunshield
8. Grenadier
9. Rifleman/Sniper with magazine-fed 7.62mm gas-operated rifle
10. Master Gunner to replace the Platoon Sergeant if he leaves the turret to operate on foot

Gavin Fighting Vehicle 4

Gavin Commander mans .50 caliber HMG and MK19 40mm Grenade Machine Gun from cupola

Driver
1. Weapons Squad leader
2. MMG Gunner A mans M240B on gunshield mount
3. MMG AG A
4. 1st Squad Team Leader B
5. AR mans M249 LMG with quick-release mounts from top troop hatch gunshield
6. Grenadier
7. Rifleman/Sniper with magazine-fed 7.62mm gas-operated rifle
8. AT Team B Gunner prepared to dismount/fire *Javelin* ATGM or *Carl Gustav* RR[10]
9. Asst AT loads *Javelins* or 84mm rounds for Gunner

All vehicles have anti-FLIR detection camouflage netting integral to their bodies
Command Gavin have C4 I means

Pre-Planned "Technology Insertions"

Advanced technologies can be applied to meet specific requirements. These include:

* Stabilized gun/sight systems incorporating day/night targeting, eye-safe laser ranging and integrated TOW controls; integrated gun and turret controls.

* Fire control systems incorporating control computers and GPS.

* Advanced fire control for helicopter engagements, including auto-tracking capabilities.

* Digital battlespace communications, integrating weapon/turret displays and controls with digital command and control communications, and combat-situational awareness. These are compatible with U.S., U.N. and NATO standards.

* Follow On To Tow (FOTT) fire/forget ATGMs utilized in TOW launchers as soon as fielded[16].

* LOSAT HyperVelocity Missiles on *Gavins* replace heavy HMMWV LOSAT as soon as fielded. HMMWV-LOSATs transfer to light Divisions[17].

*M230 30mm ASP-30 autocannon turrets with *Javelin* fire/forget ATGMs retrofitted to at least *Gavin* Commander stations when developed/fielded. May be feasible to be sized to retrofit to *Ridgways*. The low-recoil and soft-launch capabilities of the ASP-30 and Javelin allow the GC's cupola to be employed while the infantry can keep their heads out and fight situationally-aware from the top troop hatch.

Ridgway RSTA Squadron

1 Squadron of 40 RFV AMVs
30 x 3-man RFV AMV scout teams
30 x 2-man ATV scout teams

1 x Squadron CO's/XO's RFV C4 ISR AMV = 40 RFV AMVs total
33 x RFV AMVs in 3 Troops

6 x RFV 120mm mortar AMVs

60 x ATVs as supplementary scout vehicles
3 x Troop CO/XO's RFV C4 ISR AMVs = 3 RFV AMVs
3 Troops of 10 RFV AMVs = 30 RFV AMVs, 60 ATVs

2 Platoons per Troop = 10 RFV AMVs, 20 ATVs
1 Platoon Leader's RFV C4 ISR AMV = 5 *Ridgway* AMVs, 8 ATVs
2 Scout Squads per Platoon = 4 *Ridgway* AMVs, 8 ATVs
2 RFV Scout Teams A, B per Scout Squad = 2 RFV AMVs
1 ATV Scout Teams C per Scout Squad = 1 ATV

1 RFV AMV with MK-19 Grenade Machine Gun or .50 caliber HMG per AMV Scout Team
3 men per AMV scout team
2 x ATVs in Scout Team "C" moving along with "mother" RFV for forward recon work
2 men per ATV Scout Team
6 x RFV 120mm mortar AMVs
3 men per RFV 120mm mortar AMV

Equipment Common to All

2 folding All-Terrain "Mountain" Bikes available for forward scouting duties; strap to the outside of the RFV for storage.[18]

The RSTA Squadron has 6 Platoons of 5 RFV AMVs, 10 ATVs and 5-man scout teams available for reconnoitering Named Areas of Interest (NAI). Each Platoon can have a RFV 120mm mortar AMV move along for fire support, making a total of 6. 18 x UH-60L *Blackhawk* helicopters can easily move 3 Recon patrols of 6 RFV AMVs with 120mm mortars. Another option would be moving 4 Recon Patrols of 4 RFV AMVs if one of the scout carrying AMVs is substituted for a 120mm mortar *Ridgway* in one lift.

RIDGWAY Infantry Battalion

For Air Assault units within the 101st Airborne Division, the

RSTA's *Air*-Mech vehicle will be used since it can be moved by the many UH-60L helicopters available. The baseline vehicle is the *Wiesel 2* dubbed the *Ridgway* Fighting Vehicle (RFV) after the legendary World War II Airborne Hero and Commander of Troops through most of the Korean War, General Matthew B. Ridgway. The RFV can hold a Driver, Commander and 5-man dismounting infantry team. To get 3x9-man Squads in the *Ridgway* Battalions it will require 2 extra vehicles in a platoon instead of the 4 we are used to in *Bradley* and *Gavin* Battalions. Each 4-man fire-team will have its own RFV and an extra seat for the Squad Leader, Medic, Forward Observer, Radio Telephone Operator etc.

The fire support *Ridgway* Weapons Company would be created by adding lightweight laser-sight aimed 106mm RR turrets to RFV 2s to be *Ridgway* Mobile Gun Systems (R-MGS). Ridgway shooting TOW ATGMs would be *Ridgway*-Anti-Tank (*Ridgway*-AT) versions. *Ridgways* with 120mm mortars would be "*Ridgway*-120s".[19] *Ridgways* with EOCM laser targeting means would be used to transport the Sniper squad and would be called *Ridgway*-EOCMs.

RFV *Air*-Mech Vehicle 1

Ridgway Commander mans (Platoon Leader) .50 caliber HMG and MK19 40mm Grenade Machine Gun from cupola. Platoon leader with CVC helmet commands from top troop hatch, dismounts and fights the Platoon with the Weapons squad as his foot-mobile base-of-fire able to accompany them where vehicles may not.

Driver
1. RTO
2. Team Leader A
3. AR man
4. Grenadier
5. Rifleman

RFV *Air*-Mech Vehicle 2

Ridgway Commander mans .50 caliber HMG and MK19 40mm

Grenade Machine Gun from the cupola.
  Driver
  1. 1st Squad Leader
  2. Team Leader B
  3. AR man
  4. Grenadier
  5. Rifleman

RFV *Air*-Mech Vehicle 3

Ridgway Commander mans .50 caliber HMG or MK19 40mm Grenade Machine Gun from cupola.
  Driver
  1. 2d Squad Leader
  2. Team Leader A
  3. AR man
  4. Grenadier
  5. Rifleman

RFV *Air*-Mech Vehicle 4

Ridgway Commander (Platoon Sergeant) mans .50 caliber HMG or MK19 40mm Grenade Machine Gun from cupola
  Driver
  1. FO
  2. Team Leader B
  3. AR mans
  4. Grenadier
  5. Rifleman

RFV *Air*-Mech Vehicle 5

Ridgway Commander mans .50 caliber HMG or MK19 40mm Grenade Machine Gun from the cupola.
  Driver
  1. Weapons Squad leader
  2. MMG Gunner A mans M240B on gunshield mount
  3. MMG AG A
  4. 1st Squad Team Leader B

5. AR mans

RFV *Air*-Mech Vehicle 6

Ridgway Commander mans .50 caliber HMG or MK19 40mm Grenade Machine Gun from the cupola.
Driver
1. *Carl Gustav* RR or *Javelin* Gunner A
2. Asst Anti-Armor A
3. MMG Gunner B M240B
4. MMG AG B
5. Medic

RFV Mortar Section

RFV-FDC   RFV -81mm A   RFV-81mm B

RFV Weapons Company

6 x RFV-MGS (106-120mm RR)
6 x RFV-AT FOTT
6 x RFV-120mm mortars
2 X RFV-FDC
1 x RFV-EOCM Sniper

Supporting Forces from Division

9 x *Avenger* ADA HMMWVs
4 Firing teams of 2
1 C4 ISR control unit

ARTY Group

Headquarters and Headquarters Battery
Target Acquisition Battery
Service Battery
12 x M109A6 *Paladins* SP howitzers (eventually 155mm SP Howitzer *Gavins*)
6 x M777 LW 155mm towed howitzers and 6 x FMTV trucks

3 Firing batteries of 6

   9 x M270 MLRS or HIMARS (eventually UH-60L transportable T-MARs)
1 Firing battery of 8
1 C4 ISR control unit

   9 x Enhanced Fiber Optic Guided Missile (EFOGM) HMMWVs
1 Firing battery of 8
1 C4 ISR control unit

2 Engineer BFV companies
2x 12 EBFVs go to M1/M2 Battalions

1 Engineer *Gavin*-MAV Company
12 x Engineer G-MAVs go to *Gavin* Battalion

1 x Chemical *Gavin*-smokescreen Company
12 x M58 *Gavin*-smokescreen MAVs

Forward Support Battalion
___

[1] OPTEMPO concerns pushing for legislation to reduce Army deployments to 6 months:
   http://www.cnn.com/2000/US/03/06/army.duty.limits.ap/index.html
[2] Captain Marshall Miles, *Armor takes Flight, Abrams Tanks and Bradleys catch a hop into Kosovo*, (Fort Knox, Kentucky: U.S. Army Armor magazine, January-February 2000) pp. 7-12
[3] U.S. Army Infantry Center, Mechanized Infantry Battalion (Bradley Fighting Vehicle), MTOE 1999
[4] McDonnell-Douglas letter to Michael Sparks, C-17 brochure, December 5, 1994
[5] United Defense LP M113 web page describing MTVL with BFV turrets used by the Egyptian Army as the "Egyptian Fighting Vehicle"
[6] *http://www.m113.com/eifv.html*
   Egyptian Infantry Fighting Vehicle Light (EIFV) is a hybrid vehicle based on the next generation M113 chassis, featuring

the latest version of the world-class stabilized two-man *Bradley* turret - is convertible from existing assets or available as a new vehicle. It offers exceptional automotive performance, substantial armor protection and a formidable weapons suite. The vehicle is powered by a 400 hp 6V53TIA turbocharged/aftercooled diesel engine driving through an X200-4B cross drive transmission, EIFV uses many M113 common components ensuring high reliability, availability and maintainability. Applique armor provides maximum protection in a lightweight package. The vehicle carries a crew of three and six dismount Soldiers. Armed with the 25mm stabilized chaingun and a two-tube TOW launcher, this small, agile vehicle can kill either a tank or fighting vehicle. **As with all M113 variants, it is transportable on a C-130 aircraft.** Kits to convert M113s to the EIFV are available.

[6] Israeli Defense firm, Soltam informational brochures, March 2000; describing their M113-chassis based 155mm self-proppelled howitzers that are C-130 transportable. Their *Rascal* SP Howitzer with 45/52 caliber gun can range out to 41 kilometers. Their M548 SP lightweight howitzer with 39 caliber gun barrel ranges out to 30 kilometers. Both are C-130 transportable. Currently there are no C-130 transportable SP 155mm howitzers in the U.S. Army.

[7] Personal observation of Michael Sparks observing armed UH-60 Direct Action Penetrator *Blackhawks* firing 2.75 inch *Hydra-70* rockets and mini-guns at the range by Fort Bragg, North Carolina's MOUT city, April, 1995

[8] Army Technology (U.K.) OH-58D
  http://www.army-technology.com/projects/kiowa/index.html
  Tom Clancy, *Armored Cav: A Guide Tour of an Armored Cavalry Regiment*, (New York: Berkley books, 1994) pp. 143-150

[9] Natonal Guard brochure on funding prioritized Enhanced Ready Brigades (ERBs), June 1995

[10] John Pike, FAS Land Warfare Systems, Ranger Anti-Armor Weapon System (RAAWs)
  http://www.fas.org/man/dod-101/sys/land/m3-maws.htm

[11] Stan Crist, *Too Late the XM8?*, (Fort Knox, Kentucky: U.S. Army Armor magazine, January-February 1997), pp. 16-

19 Bofors enhanced 106mm RR 3A-HEAT-AT details

[12] United Defense LP M8 Armored Gun System brochure, San Jose, California: 1999.

Tom Clancy, *Armored Cav: A Guide Tour of an Armored Cavalry Regiment*, (New York: Berkley books, 1994) pp. 94-96

John Pike, FAS Land Warfare Systems, 105mm tank gun ammunition: *http://www.fas.org/man/dod-101/sys/land/105t.htm*

[13] Raytheon Elevated Tow System press release: *http://www.raytheon.com/press/1999/aug/elevatow.html*

[14] Christopher Foss, *Mortars target the Middle East*, (London: Jane's Defense Weekly, 23 July, 1994) pp. 22-23

[15] U.S. Army, Washington DC: *Operator's Manual 7.62mm M24 Sniper Weapon System (SWS)* NSN 1005-01-240-2136, TM 9-1005-306-10, , June 1989

[16] John Pike, FAS Land Warfare Systems, Follow-On-To-Tow: *http://www.fas.org/man/dod-101/sys/land/fott.htm*

[17] Lockheed-Martin LOSAT brochure, Austin, Texas: June 1999

[18] Wade Shaddy, *Operation Dark Claw*, (Emmaus, Pennsylvania: Bicycling magazine, September 1999) pp. 20-22

[19] Mak *Wiesel 2 with 120mm mortar* brochure, Kiel, Germany: 1999. *Wiesel 2* with 120mm mortar is fired after hydraulically lowering supports to the ground.

Army Combat Division: HEAVY

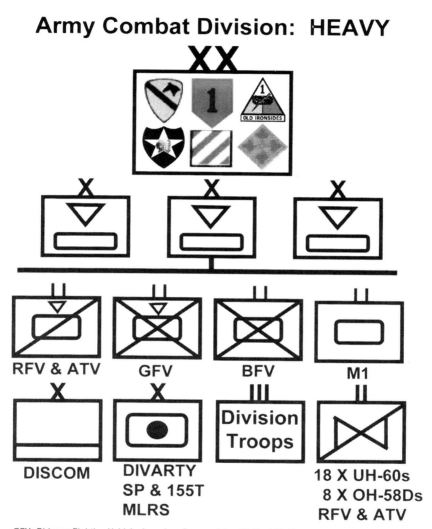

RFV: *Ridgway* Fighting Vehicle; based on German 4-ton *Air*-Mech Vehicle (UH-60L Lift Capable)
GFV: *Gavin* Fighting Vehicle; based on 11-ton M113A3 Armored Carrier (CH-47F Lift Capable)
MLRS: Track-mounted multiple rocket system (C-5/C-17 lift capable)
Paladin M109A6: 155mm self-propelled howitzer (C-5/C-17 lift capable) , M777 LW towed 155mm

Army Combat Division: HEAVY
(Major Charles Jarnot and Mike Sparks)

Army Combat Division: LIGHT

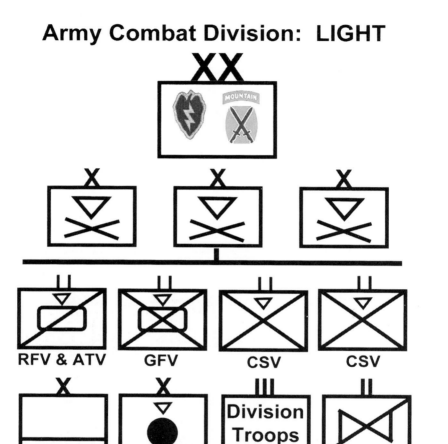

RFV: *Ridgway* Fighting Vehicle; based on German 4-ton *Air*-Mech Vehicle (UH-60L Lift Capable)
GFV: *Gavin* Fighting Vehicle; based on 11-ton M113A3 Armored Carrier (CH-47F Lift Capable)
HIMARS: 5-ton FMTV truck-based MLRS rocket system (CH-47F lift-capable)
T-MARS: Proposed MLRS Trailer Mounted Rocket Pod (UH-60L Lift Capable)
Lt-Wt 155: M777 Lightweight Towed 155mm Howitzer (UH-60L Lift Capable)

Army Combat Division: LIGHT
(Major Charles Jarnot and Mike Sparks)

Army Combat Division: Air Assault

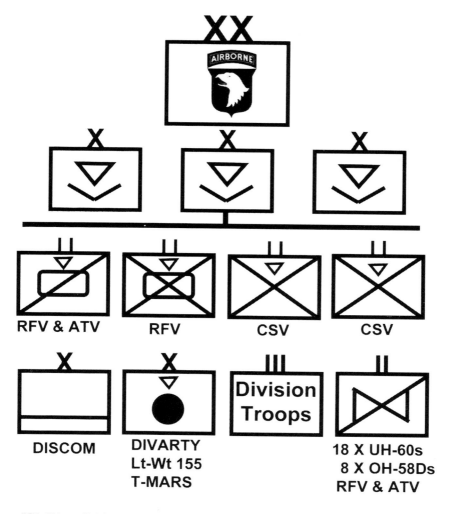

RFV: *Ridgway* Fighting Vehicle; based on German 4-ton *Air*-Mech Vehicle (UH-60L Lift Capable)
GFV: *Gavin* Fighting Vehicle; Based on 11-ton M113A3 Armored Carrier (CH-47F Lift Capable)
T-MARS: Proposed MLRS Trailer Mounted Rocket Pod (UH-60L Lift Capable)
Lt-Wt 155: Light Weight Towed 155 mm Howitzer (UH-60L Lift Capable)

Army Combat Division: AIR ASSAULT
(Major Charles Jarnot and Mike Sparks)

Army Combat Division: Airborne

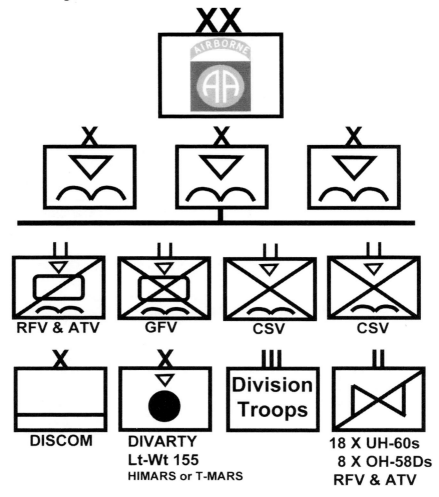

CSV: Combat Support Vehicle; based on Commercial 6 X 6 ATV, 3000 lbs payload.
RFV: *Ridgway* Fighting Vehicle; based on German 4-ton *Air*-Mech Vehicle (UH-60L Lift Capable)
GFV: *Gavin* Fighting Vehicle; based on 11-ton M113A3 Armored Carrier (CH-47F Lift Capable)
HIMARS: 5-ton FMTV truck-based MLRS rocket system (CH-47F lift-capable)
T-MARS: Proposed MLRS Trailer Mounted Rocket Pod (UH-60L Lift Capable)
Lt-Wt 155: M777 Lightweight Towed 155mm Howitzer (UH-60L Lift Capable)

Army Combat Division: AIRBORNE
(Major Charles Jarnot and Mike Sparks)

Gavin Fighting Vehicle Infantry Battalion

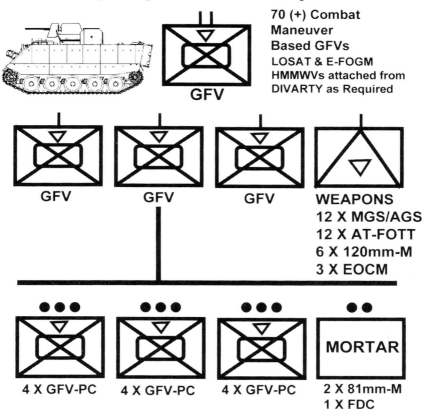

70 (+) Combat Maneuver Based GFVs
LOSAT & E-FOGM
HMMWVs attached from DIVARTY as Required

GFV

GFV   GFV   GFV   WEAPONS
                     12 X MGS/AGS
                     12 X AT-FOTT
                     6 X 120mm-M
                     3 X EOCM

4 X GFV-PC   4 X GFV-PC   4 X GFV-PC   MORTAR
                                          2 X 81mm-M
                                          1 X FDC

GFV-PC: *Gavin* Fighting Vehicle; Based on 11-ton M113A3 Armored Carrier (CH-47F Lift Capable) Equipped with a M-230 30 mm autocannon, *Javelin* ATGM & FLIR.
GFV-MGS: Mobile Gun System, 90-105mm low velocity gun or 106mm Recoilless Rifle based on GFV
M8 AGS: 105mm high-velocity shoot-on-the-move gun; 17-ton light tank; parts from other Army vehicles
GFV-AT: Elevated Tow System (ETS) shoots fire/forget, self-guiding, top-attack ATGMs in defilade
EOCM: Counter-Sniper Squad vehicles with electro-optical counter-measures
LOSAT: Line Of Sight Anti-Tank, Hyper-Velocity Rockets, 5 Km Range, 1600 meter/second SABOT.
E-FOGM: Enhanced range Fiber Optic Guided Missile, 15 Km over horizon range, top-attack.
120mm: Mortar capable of firing Precision Munitions, range 7 Km, top attack Anti-Tank.
Sensors: Sensor platoon at HHC with one GFV with trailer, launch up to 12 X 200lb UAVs.

NOTE: Additional *Gavin*/HMMWVs in HHC to support TOC, Medical, Maintenance and Resupply

Gavin Fighting Vehicle Infantry Battalion
(Major Charles Jarnot and Mike Sparks)

Ridgway Fighting Vehicle Infantry Battalion

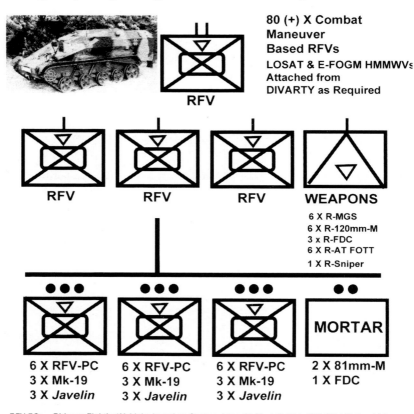

80 (+) X Combat Maneuver Based RFVs
LOSAT & E-FOGM HMMWVs Attached from DIVARTY as Required

WEAPONS
6 X R-MGS
6 X R-120mm-M
3 x R-FDC
6 X R-AT FOTT
1 X R-Sniper

6 X RFV-PC
3 X Mk-19
3 X *Javelin*

6 X RFV-PC
3 X Mk-19
3 X *Javelin*

6 X RFV-PC
3 X Mk-19
3 X *Javelin*

MORTAR
2 X 81mm-M
1 X FDC

RFV-PC: *Ridgway* Fighting Vehicle; based on German 4-ton *Air*-Mech Vehicle (UH-60L Lift Capable) Equipped with a Mk-19 40mm Auto Grenade Launcher or *Javelin* Fire/Forget ATGM & FLIR.
LOSAT: Line Of Sight Anti-Tank, Hyper-Velocity Rockets, 5 Km Range, 1600 meter/second SABOT.
E-FOGM: Enhanced range Fiber Optic Guided Missile, 15 Km over horizon range, top attack.
R-120mm: Mortar capable of firing Precision Munitions, range 7 Km, top-attack Anti-Tank.
Sensors: Sensor section at HHC with one RFV with trailer, launch up to 2 X 200lb UAVs.

NOTE: Additional RFVs/HMMWVs in HHC to support TOC, Medical, Maintenance and Resupply

Ridgway Fighting Vehicle Infantry Battalion
(Major Charles Jarnot and Mike Sparks)

Ridgway 120mm mortar variant (Mak-Bob Novogratz)

Kosovo situation map: What if we had attacked in 1999?
(Major Charles Jarnot and Mike Sparks)

Chapter 9

Air-Mech-Strike *Phalanx* in Action

"In our time this means the recognition that ground warfare has become three-dimensional and that Air-mobility is a critical factor on the battlefield. The infantryman of future fights will take the high ground by Air Assault."
—General John R. Galvin (USA, Retired), former NATO Commander, *Air Assault: The development of Air-mobile Warfare*

Kosovo, what If we had attacked?
"'A clever military leader will succeed in many cases in choosing defensive positions of such an offensive nature from the strategic point of view that the enemy is compelled to attack us in them."
—Moltke[1]

June 10, 1999
D + 20

NATO's air campaign to get the Serbian war criminal Milosevic to stop the ethnic cleansing of Kosovo province is not working as planned. The Serbs fully understand the U.S. preferred use of Precision Guided Munitions and other stand-off tactics are based on our aversion to U.S. casualties. Learning from Saddam Hussein in *Desert Storm*, and the previous NATO campaign in Bosnia, the Serb Army implements asymmetric tactics to negate the firepower-attrition strategy of the U.S.[2]

The Serbs deploy a classic counter-firepower defense dispersing armor units throughout restricted terrain, and villages. While the Serb Army disperses, special police and paramilitary forces continue to commit atrocities throughout Kosovo.[3] Milosevic deploys an aggressive and thorough information warfare campaign further degrading public support for NATO and the U.S.

The NATO Atlantic Counsel remains firm in accomplishing their stated objectives of ending atrocities, and ending Milosevic's reign of terror. NATO Command Headquarters propose and the national governments approve ground force operations along with a continued air campaign. Permission granted to deploy additional U.S. Forces.

June 20, 1999 Force Disposition:

The majority of the European NATO forces are already in position in Macedonia; British and German forces are *Air*-Mech-Strike capable with *Scimitar* and *Wiesel* type AFVs. U.S. forces continue to build up in Macedonia and Albania for either peace support or combat operations. The 2^d Brigade of the 1st Infantry Division ("*Big Red One*") draws pre-positioned war stocks from southern Europe and then moves to Albania by barge; their 3d Brigade has already been moving by Roll-on/Roll-off ship (RO/RO) to a Sea Port Of Debarkation (SPOD) in Italy. The 3d Brigade barges to Albania; all Brigades of the *Big Red One* are *Air*-Mech-Strike capable.

The 4th Marine Expeditionary Brigade (MEB) is positioned in the Adriatic Sea to reinforce the U.S. Army's 5th Corps in Albania.

The Division Ready Brigade of the 82d Airborne, positioned in Fort Bragg, North Carolina and the 1st Battalion of the 508th in Vicenza, Italy assemble at Ramstein Air Force Base, Germany for future combat operations. They are *Air*-Mech-Strike capable.

A Brigade from the 101st Air Assault Division, Fort Campbell Kentucky, begins deployment to Italy. 1st Brigade at Fort Riley, Kansas leaves Beaumont, Texas by RO/RO ships that will also offload at the SPOD in Italy to transload by barge to Albania.

15 July

Rehearsals and final preparations for combat are finalized. NATO commanders report to either U.S. Army V Corps or the British Rapid Reaction Corps (RRC) Commanding Generals.[4] The Commanding Generals report to LANDCENT that they are "*ready for combat*". From the beginning of the NATO air campaign, extensive efforts to improve the poor roads/bridges of the Albania/Macedonia area of operations have been underway around-the-clock. These actions were taken to improve humanitarian relief. If an attack is ordered, interior lines of resupply will be in place.

Political bickering comes to closure due to continued killing, the massive flow of refugees, and the anticipation of the coming winter. Ammunition and fuel stocks were built up for combat operations as humanitarian supplies continued to be brought in to take care of the refugees.

20 July 0700 Hours
NATO gives the "Green Light"

The LANDCENT Commander positioned in Macedonia receives briefbacks from both his Corps commanders. The 5th Corps Commander will have the 1st Infantry Division as the main effort attacking from Albania into Kosovo with a supporting attack in the north by the 4th MEB.

The Rapid Reaction Corps will conduct a supporting attack from Macedonia north into Kosovo consisting of the German *Wiesel* Airborne Brigade, the British *Air*-Mobile Brigade with *Scimitar* light tanks and *SupaCat* ATVs, and the Italian *Folgore* Paratroopers with *M113s*. Six additional European Brigades will follow and support.

2200

NATO air strikes and Army long-range rocket fire from Task Force *Hawk* in Tirana, Albania concentrate on all suspected Serb C2; power grids, LOCs and armored forces to "stun" the enemy as the Rapid Reaction Corps (RRC) begins their supporting attack.

The RRC advances slowly into Kosovo due to Serb artillery fire and extensive mines encountered through the narrow mountainous defiles.

The RRC employs their *Air*-Mech-Strike elements to clear the ridgelines on either side of the defiles supporting the heavy armor that must pass through the choke points.

Progress is slow but the 1st echelon of the Serb Army is being forced to displace by RCC *Air*-Mech-Strike maneuver.

2359

The 4th MEB conducts a feint into Montenegro, then lands in northern Albania, moves east to fix Serb units on the Kosovo border

21 July 99
0100

The 82d Airborne is 20 minutes out from their Drop Zones, having left Ramstein 3 hours prior, to parachute vicinity Pristina, secure the airfield for future reinforcements, and cut the Serb Army Lines Of Communications north, east and south of this key terrain.

0200

1st Infantry Division is in attack positions along Albania/Kosovo border. Their *Gavin* Battalions have secured infiltration routes not suited for the heavy *Abrams* tanks or the *Bradley* Fighting Vehicles. Reports continue to come in from the RSTA units in *Ridgways*, supported by small Kosovo Liberation Army/U.S. Army Special Forces teams, who provide local intelligence on enemy positions and minefields.[5]

0215

Robust SEAD fires are shifted from the Airborne operation to the Air Assaults supporting the 1st Infantry Division's main effort.

0230

2 Battalions of the 1st Brigade of the 101st Air Assault Division land behind Serb defenses in Kosovo near Prizren, cutting LOCs, north, south, east and west…

Throughout the area of operations, Paratroopers, Air Assault Soldiers and RSTA scouts continue to employ observed fires on enemy forces.

0300

The heavy battalions of the 1st Infantry Division cross the Line of Departure (LD) and fight through the mountainous defiles into Kosovo supported on the flanks by *Gavin* battalions.

One battalion of the 1st Brigade 101st remains the Division reserve in Tirana with TF *Hawk*. AH-64s from TF *Hawk*, in support of the 82d Airborne with deep attacks destroy Serb T-84 tanks on LOCs vicinity Pristina.

0530 Dawn over Kosovo

The Rapid Reaction Corps continues to encounter bitter fighting through the densest array of obstacles pushing to link up with their *Air*-Mech-Strike elements.

1st Infantry Division is 5 kilometers from Prizren moving slow due to minefields, but their *Gavin* Battalions have linked up with their helicopter-inserted RSTA Squadrons.

The 4th MEB has fixed Serb forces north of the 1st Infantry Division along the border.

Airborne forces have secured their objectives around Pristina continuing to interdict forces moving into Serbia.

NATO units catch many Serb forces in the daylight emerging from hide sites where they are destroyed or captured.

Many shocked Serbs surrender to NATO forces thanks to an aggressive Information Operations (IO) campaign integrated with the 3-Dimensional attacks, which shatter the will of Serb Army in Kosovo.

30 July

Except for the mopping up small Serb elements cut off in the mountains in Kosovo, combat operations have ended and the consolidation phase has begun.

Major Theater War: Korea *Air*-Mech-Strike in Far East Asia

"Before upsetting the enemy's equilibrium by one's attack, it must be upset before the real attack in launched. A general in not justified in launching his troops to a direct attack upon an enemy firmly in position."
—B.H. Liddell-Hart, *Strategy*, page 147

Prelude to Conflict: After several years of disastrous economic conditions in the People's Democratic Republic of Korea (North Korea), on March 1, 2000 following the winter military training cycle, North Korea initiated a fierce volley of political accusations against their neighbor to the south and the United States. The Republic of Korea and her ally the United States, have witnessed these irrational attacks before and has carefully monitored the military activity north of the border. Often these tantrums from the North Korean dictator Kim Jung-IL is followed by some military preparation for invasion such as troop movements, vehicle marshalling and increased aircraft and communication activity. A recognized long term goal of North Korea is to dull the sensitivity of the defending forces to the south through repetition, thereby reducing the window of unambiguous warning of an actual attack.

Currently, the two nations are still locked in a tense military confrontation, which has taken a turn for the worse in the last decade. The South Korean economy has boomed under a democratic government and its application of free market economics. In contrast, the police state in the north has been on a steep economic downward spiral since the Soviet Union collapsed and the massive Russian economic aid dried up. The Communist Chinese, angered over the North Korean's alliance with their socialist rivals in the Soviet Union has not replaced the latter's aid to anywhere near the extent that it was under the USSR The People's Republic of China (PRC), uses its limited support of North Korea in a quid-pro-quo detante' exercise with the United States over its support of Taiwan. With no real hope of expecting the return of Russian aid, many world economic experts familiar with the country predict a collapse in the near future.

A Major Theater War (MTW) with a desperate, collapsing North Korea is unfortunately, a very distinct possibility. The border between these two countries is the most heavily defended frontier in the world. To the North, lies a million plus man army deployed in a Soviet-style offensive disposition. Supporting these mechanized and truck-mobile formations is the world's largest concentration of heavy artillery and rocket batteries, capable of delivering vast amounts of chemical munitions. Large numbers of special operations units also ready to support a planned invasion using tunnels, submarines, light aircraft and commando boats to infiltrate into the rear of defending South Korean and American troops.[6] The disaster of a war in Korea would undoubtedly result in the massive loss of life, the likes of which have not been seen since the Second World War. *Air*-Mech-Strike offers the best solution to outflanking the massive fortifications and minefields and thus avoids the large casualties associated with close combat in such an environment.

Phase One; the Invasion: On March 1, 2003, following the winter military training cycle, the North Korean Army attacks South Korea. The invasion of the south is initiated by a massive artillery, rocket and missile barrage that includes large amounts of persistent chemical agents. The effects of the massive use of chemical warfare create the operational opportunity that the North Korean Army was hoping for. A slow down in the deployment of air and land forces from the United States caused by the huge backlog of casualties and the chemical contamination of South Korean ports and airfields has provided the attacking North Korean Corps a "window" of time in which press home their advance southward.

The allied Commander and Chief of Combined Forces Command (CINC-CFC) now needs a method of quickly blocking a penetration down the valley paths in the mountainous terrain leading to the southern port of Pusan. The CFC firepower from Naval, Air Force and artillery systems are seriously attriting the North Korean Corps, but their willingness to accept large casualties in exchange for advances southward has paced them for a chance at success. The U.S. 2nd Infantry

Division, already deployed in Korea has accepted its third Brigade from the Continental United States (CONUS) and is committed into the fray. For the next three to four weeks, when ships arrive from the United States carrying additional Divisions, the CINC-CFC has no additional maneuver forces to act as an operational reserve to quickly maneuver in the mountainous Korean terrain to stop a North Korean Corps threatening to break out to Pusan.

Phase Two: NCA Commits the Air-Mech Option: With the war only ten days old, and the situation becoming critical, the National Command Authority, (NCA) authorizes the commitment of the 101st *Air*-Mech option and the Army Prepositioned Commercial Aircraft (APCA) fleet necessary to transport the force as leaders of a U.N. counter-offensive. Already alerted since the start of the conflict, the recently retired airline pilots, on a flight-qualified retainer, have assembled at the APCA airfields located in New Mexico and Arizona. The 30 Boeing 747s of the APCA are maintained in fly-away condition by a contractor and preloaded with Combat Support Vehicles and 300 *Ridgway* Fighting Vehicle 4 -ton *Air*-Mech vehicles to meet contingencies ranging from closed-terrain, jungle warfare to open-terrain desert-scenario Brigade Task Force's requirements with a 24 UH-60L helicopter Battalion along with 72 hours of vehicle fuel and ammunition. The Army secured these older Boeing 747s and some assorted cargo-capable DC-10 and MD-11 wide body jets from the stock of semi-retired aircraft held by the major airlines. These aircraft are normally older airframes that are no longer economical to operate due to the age of their engines and key systems. Contractors will normally maintain these aircraft for the major carriers until a foreign buyer can be located or seasonal demands require their service again. The Army had the contractor modify the 30 leased airliners with nose opening cargo doors, strengthened floors and cargo/pallet floor runners along with a military communication and a GPS navigation suite.

The APCA departs with 24 hours of notice to Osaka International Airport in Japan where they are met by some 3000 troops from the Division Ready Brigade Task Force on alert with the 101st Airborne Division, who were simultaneously flown in on six Boeing 747s. The helicopters are reassembled and the flown to the Theater Support Base in Kitakyushu Japan, only some 225 kilometers from Pusan South Korea. The Division Ready Brigade Commander reports fully operational at the TSB within 96 hours of deployment order.

The CINC-CFC see the first indications of one of the three main attacking North Korean Corps breaking down the defenses of the combined U.S. and ROK Army defenders. The CINC-CFC commits the 101st Division Ready Brigade from Japan. The *Air*-Mech Brigade uses its three-dimensional capability to air move its Battalions from Japan immediately behind a key defending ROK division using its UH-60 Battalion. The helicopters are able to fly 16 to 18 hours each per day due to the *Air*-Mech-Strike force model of two crews per aircraft. The net effect is the one deployed helo unit generates twice the daily sortie rate of the older, single-crew manned force model. The RFVs fitted as Streamlined External Loads (SEL) closely under the fuselage of the UH-60 *Blackhawk*, equipped with stork-like landing gear extensions, fly in the battle area without the additional risk of conventional sling-loading via their slow speeds and high above-ground altitudes.

The 100 *Ridgway* Fighting Vehicles deployed from the relative safety of Japan provided exactly the responsiveness for an operational reserve that could quickly traverse the mountainous terrain of Korea with air assault speeds yet retain a credible mechanized maneuver capability upon arrival. The RFVs crews and infantry maneuvered into defensive positions under armor protection and in vehicles offering over-pressurization protection from chemical agents.[7] The Brigade brings immediate firepower to the close battle in the form of *Javelin* anti-tank missiles, MK-19 Grenade Machine Guns, M-230 30mm autocannons and 120mm heavy mortars. The fresh array of helicopter mounted Unmanned Aerial Vehicles (UAVs) available down to the Company Commander also boosts the situational awareness array of the Brigade's gaining higher headquarters.

At the same time that the Division Ready Brigade of the 101st Airborne Division (AA) was deploying from CONUS, 2nd

Brigade, the first up Division Ready Brigade (DRB) of the 82nd Airborne Division, loaded on to strategic lift now available from a shift in force flow priority brought about by the criticality of the situation on the peninsula arrives in Alaska to link up with the 1st Battalion of the 501st "*Geronimos*" and the rest of the 172d "*Arctic*" Brigade. The rapid advance of the North Korea Corps presented the CINC-CFC with an opportunity to attack an exposed flank of one of the Corps. Such an attack would have to be at least executed at Brigade strength and have to have the tactical mobility, protection and firepower to compete against enemy mechanized and truck mobile units. The 2nd Brigade and the 172d Brigade from Alaska execute a *Paramech* assault with their *Gavin* Fighting Vehicles parachute dropped by USAF C-17s into the flank of the North Korea 806th Infantry Corps. An Assault Zone (combination drop and landing zone) is set up for the 172d Brigade to airland using USAF and U.S. Air National Guard C-130H/Js.

The responsiveness of these two *Air*-Mechanized Airborne and Air Assault organizations gained the edge needed by the CINC-CFC to regain the initiative. Current model Airborne and Air Assault force structures would have been less viable assets as the CINC-CFC's operational reserves. Their dismounted infantry would lack the tactical mobility, protection, sustainment and firepower to effectively stall the advance of a typical North Korean truck-mobile or mechanized infantry unit.

In the follow-on, counter-offensive some 45 to 60 days later, the now full-up 101st and 82nd Divisions offer the CINC-CFC an operational vertical envelopment of North Korea. These *Air*-Mech capable units can insert GFV and RFVs deep via both helicopters and para-drops to cut off the lines of communication to the People's Republic of China promoting a speedy collapse of the North Korean Government. Airlanded and paradropped fuel pods[8] for the helicopters along with the ability to mid-air refuel and top-off on Navy ships sustains the deep *Air*-Mech-Strike assault.

Kuwait: *Air*-Mech-Strike Deterrence Scenario

"Appear at points which the enemy must hasten to defend, march swiftly to places where you are not expected"
—Sun Tzu, *The Art of War*

Prelude to Conflict: With over half of the world's energy needs coming from the oil in the Persian Gulf, America will consider the free flow of petroleum to be a National Vital Interest for the foreseeable future. The Army has invested in propositioning at least two Brigade sets of equipment in Kuwait as deterrence to either Iraq, Iran or both. In this scenario both countries are "*saber rattling*" by beginning a series of coordinated troop deployments and conducting "*scheduled military exercises*" close to the border with Kuwait.

Deterrence Deployment: Upon indications of stepped up military activity in Iran and Iraq, the National Command Authority orders the deployment of a Brigade from the 10th Mountain Division and the 3rd Infantry Division. The Soldiers are flown in and draw pre-positioned Brigade stocks of equipment. However, the situation escalates and the "*training maneuvers*" begin to look like a prelude to an attempt to invade Kuwait by a coalition of Iraq and Iran. Live ammunition has been issued to Iraqi and Iranian armored formations at the same time in their respective countries.

Up-Grade the Deterrence with an *Air*-Mech Brigade: Based on the new indicators, the National Command Authorities (NCA) authorized the release of the Army Preposition Commercial Aircraft fleet to support the deployment of the Divisional Ready Brigade Task Force from the 101st Airborne (AA) Division. The preloaded wide-body jets begin the cross globe departures for Kuwait within 24 hours of the NCA's decision. Both Brigade equipment and the deploying troops meet up in Kuwait and within 96 hours the all RFV-equipped *Air*-Mech-Strike Brigade Task Force is operational.

Long Range *Air*-Mech-Strike: The deployment of the *Air*-

Mech-Strike Brigade gives the tactical Commander in Kuwait the ability to strike at the flanks of either the Iraqi or Iranian columns well forward of the Kuwait border should a pre-emptive spoiling attack be authorized. Both columns could be interdicted at the numerous choke points in the area using Battalion-size Task Forces. This capability will cause the enemy to reconsider their course of actions and either makes them act with greater deliberation or cause them to deploy security forces in these regions to counter the *Air*-Mech-Strike threat.

***Air*-Mech-Strike Corrects "*Speed Bump*" Paradigm:** The results of having an air-insertable force with the ability to move at mechanized speeds under armor protection greatly enhances the aggregate deterrence value of a rapid deployment early-entry force. The enemy must weigh their value when considering their course of action. The current model of rapidly deploying a Brigade that is 100 dismounted amounts to presenting a potential adversary with the fatalistic "*Speed Bump*" that is easily bypassed in a wide-open desert environment.[9] In contrast, AMS forces can deploy by air to the conflict area and can execute a mobile, area defense with superior *Air*-Mech mobility than that of the enemy.

---

[1] Joint Readiness Training Center Observer/Controller Heavy/Light Forces Briefing paper, 1995
  [2] Serb asymmetric tactics
http://news.bbc.co.uk/low/english/special_report/1998/kosovo/newsid_337000/337679.sm
  [3] Serb Army atrocities in Kosovo:
http://www.hrw.org/reports/2000/fry/
http://www.hrw.org/hrw/reports98/kosovo/
http://www.hrw.org/hrw/reports/1999/glogovac/
This BBC site is great for eye witness accounts:
http://news.bbc.co.uk/hi/english/static/inside_kosovo/default.stm
  [4] NATO Rapid Reaction Corps:
http://www.nato.int/docu/handbook/hb32606e.htm
    http://www.defenselink.mil/pubs/allied_contrib99/rs99-chpt3.html

http://call.army.mil/call/fmso/fmsopubs/issues/ifor/chpt2.htm
http://www.arrcmedia.com/
http://www.dfait-maeci.gc.ca/english/news/newsletr/un/rap7.htm

[5] U.S. Army Special Forces targeting in Kosovo
http://www.janes.com/defence/features/kosovo/special_forces.html

[6] North Korean People's Army:
http://www.fas.org/nuke/guide/dprk/agency/army.htm
http://www.fas.org/nuke/guide/dprk/agency/kpa-guide/index.html

[7] NBC protection is a standard option for the Mak *Wiesel* family of AFVs.

[8] U.S. Army and Air Force, *Airdrop of Supplies and Equipment: Rigging Forward Area Refuelling Equipment (FARE)*, (Washington D.C: Government Printing Office) 28 February 1983
http://155.217.58.58/cgi-bin/atdl.dll/fm/10-537/fm10-537.htm

[9] David H. Lippman, *Desert Dawn: North Africa Before Rommel*
(Italians in North Africa in WWII)
http://www.magweb.com/sample/seuropa/seu55daw.htm

But behind the numbers and glittering Fascist regalia lurked serious weaknesses that Graziani himself knew. The Italian 10th and 5th Armies in Libya marched on foot, while the British rode in trucks... Just as importantly, the Italian forces had poor equipment. Armored cars dated back to 1909. The L3 tank only mounted two machine guns. The under-powered and thinly-armored M11 was little better — its 37mm gun could not traverse. The heavyweight M13 packed a 47mm gun, but crawled along at nine miles per hour. None could match the [tracked] British *Matilda* with its 50mm armor and 40mm gun. Italian troops were short of anti-tank guns, antiaircraft guns, ammunition, and radio sets. Artillery was light and ancient.

Korean situation map: DEFENSIVE operations 2003
(Major Charles Jarnot and Mike Sparks)

Korean situation map: OFFENSIVE operations 2003
(Major Charles Jarnot and Mike Sparks)

KUWAIT *Air*-Mech-Strike Scenario Circa 2003 DETERENCE

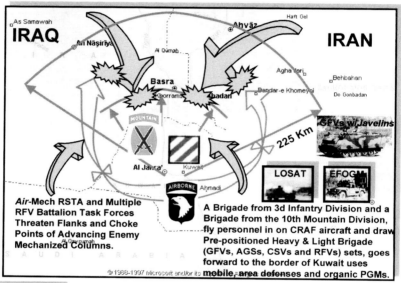

Air-Mech RSTA and Multiple RFV Battalion Task Forces Threaten Flanks and Choke Points of Advancing Enemy Mechanized Columns.

A Brigade from 3d Infantry Division and a Brigade from the 10th Mountain Division, fly personnel in on CRAF aircraft and draw Pre-positioned Heavy & Light Brigade (GFVs, AGSs, CSVs and RFVs) sets, goes forward to the border of Kuwait uses mobile, area defenses and organic PGMs.

101st Airborne (AA) Division and 3d Infantry Division Projects from CONUS an *Air*-Mech-Strike Brigade (+) Heavy Task Force, (200 RFVs, 32 Helos, 14 M1A2s, 14 M2A3 BFVs (some with Stingray EOCM), 4 MLRS and 2 Patriot Firing Units). Use of APCA, 30 Boeing 747 sorties, frees 30 C-17 sorties for the Outsized Weapon Packages. Theater Support Base = AD DAMMAN (Saudi Arabia)

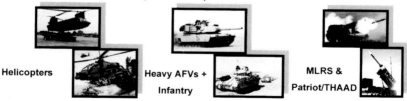

Helicopters   Heavy AFVs + Infantry   MLRS & Patriot/THAAD

Kuwait situation map: DETERRENCE operations 2003
(Major Charles Jarnot and Mike Sparks)

Wing-In-Ground Effect Aircraft

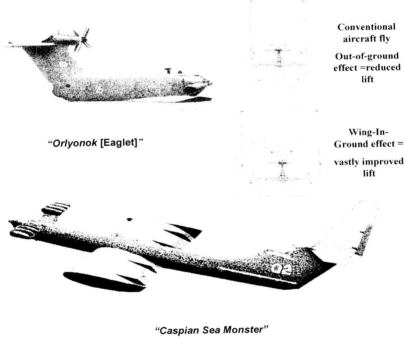

"*Orlyonok* [Eaglet]"

Conventional aircraft fly Out-of-ground effect =reduced lift

Wing-In-Ground effect = vastly improved lift

"*Caspian Sea Monster*"

(Illustrations by John Richards)

Russian Wing-In-Ground (WIG) effect aircraft (John Richards)

Chapter 10

Air-Mech-Strike Training and Readiness

"The American Soldier is a proud one and he demands professional competence in his leaders. In battle, he wants to know that the job is going to be done right, with no unnecessary casualties. The noncommissioned officer wearing the chevron is supposed to be the best Soldier in the platoon, and he is supposed to know how to perform all the duties expected of him. The American Soldier expects his Sergeant to be able to teach him how to do his job. And he expects even more from his officers."
—General Omar N. Bradley

Joint Publication 3-0 *Doctrine for Joint Operations* states; "*the U.S. military must have trained and ready forces that are rapidly and strategically deployable; and initially self-sufficient for response to spontaneous, un-predictable crisis*".[1] Unit training must focus on the readiness to deploy anywhere in the world and conduct "*mission essential tasks*" throughout the spectrum of conflict. Forces must train and prepare themselves for self-sufficiency for as long as possible with the mobility to move fast and hit hard. Units must have the ability to task organize quickly, tailored to respond to diverse and contradictory environments.[2] *Air*-Mech-Strike forces must be trained and ready not only to fight, but also to marshal and move on short notice. Units must conduct Emergency Deployment Readiness Exercises (EDREs) to rehearse strategic deployment and tactical mobility.

The ability to move in and over all types of terrain is critical to *Air*-Mech-Strike responsiveness and tactical agility. To move in all situations means mobility, and the supreme tactical principle for *Air*-Mech-Strike is mobility. We enhance mobility when supported by surprise and leader initiative, displaying quick decisions and quick movements, doing what the enemy does not expect, and employing multiple means and methods.[3]

The key to *Air*-Mech-Strike readiness, are well-trained, quality people, led by confident NCOs and officers, that understand the human dimension of combat; and the doctrine, tactics, techniques, and procedures to employ this concept in combat. Junior leaders must have the ability to operate effectively amid the chaos of battle, to generate chaotic situations for the enemy, and to be positioned and prepared to take advantage of the opportunities a chaotic battlefield presents.[4]

Successful maneuver in such difficult circumstances demains an elasticity of mind. Training must focus on how to thrive in uncertainty and ambiguity.[5] *Air*-Mech-Strike leaders are taught to act independently, to anticipate second and third order effects from their actions and the actions of others, the necessity of situational awareness to shape the battlefield for combat and operations other than war, and to take advantage of all resources simultaneously. Key to *Air*-Mech-Strike is the flexibility offered for a leader to achieve positional advantage over the enemy using effects of terrain, creating vulnerabilities to his battlefield operating systems, controlling his direction of movement, degrading his decision cycle, and creating favorable force ratios for close combat. Our leaders' ability to use this flexibility, along with the "fifth element" of combat power – Information Operations (IO), provides the combat agility our commanders need in uncertain, stressful situations. Understanding how to employ IO: civil affairs, psychological operations, public affairs, EW, and direct action; protection of friendly info-systems; and the use of relevant information and intelligence provides the commander additional combat multipliers. Our proficient use of IO is critical when dealing with what Lieutenant Colonel Ralph Peters (Retired) describes as "*the new warrior class*". The commander's ability to synchronize IO with the agility offered by *Air*-Mech-Strike enhances our success in operations against symmetric or asymmetric threats.

Army leaders must understand how to fight the *Air*-Mech-Strike 2-D/3-D maneuver concept. Leadership competence is the most critical factor influencing unit morale and tactical proficiency. After *Air*-Mech-Strike concepts and doctrine are developed, our Divisions, Brigades and Battalions will transform to newly organized, equipped and manned units. New Tactics,

Techniques and Procedures (TTP) will be taught to our leaders who will in turn train their Soldiers.

Genghis Khan refined his force into heavy and light cavalry—that complimented each other tactically.[6] The Divisional, Brigade and Battalion organization in *Air*-Mech-Strike is all about forces that compliment each other—optimum combined arms teams. The goal of leaders is to get their Soldiers to think and fight as one team, unified in purpose. Solidarity and confidence cannot be improvised; they have to be built over time. They are born of mutual acquaintanceships which establishes pride and develops unity. And, from unity comes in time the feeling of force, that force which gives the attack the courage and confidence of victory.[7]

Training will focus on the combined-arms team, jointness, and the elements of combat power: maneuver, firepower, protect the force and leadership. Leaders must understand *Air*-Mech-Strike force capabilities to reconnoiter, conduct decisive maneuver, fix, destroy and exploit the enemy in all terrain, environments and throughout the range of operations. Besides knowing the system capabilities, commanders must become experts in fitting the parts together on the battlefield to produce the combat power synergy to destroy the enemy's will to fight. Leaders must be taught how to focus combat power and shift that power as the situation changes. They must know what parts have the agility to rapidly adjust to new battlefield dynamics, and the mobility factors essential to getting the right units to the right place at the right time. Commanders must understand the cumulative effect relationship of physical destruction and the influence on the moral domain, as a result of *Air*-Mech-Strike operations. Leaders will train to take away the energy of our enemy's forces and the heart of their Generals.[8]

Teamwork is a priority in the *Air*-Mech-Strike combined-arms team. As Colonel Dan Malone states in "Small Unit Leadership"; "the FIRST two letters of U.S. Army are (US), and the LAST two letters are (MY). With the multi-functional mix of combat, combat support and combat service support forces in *Air*-Mech-Strike, teamwork is the "glue" that holds the capability together.

The secret of successful war lies in keeping men in condition of mental alertness and physical well-being which insures that they can and will move when given a competent order.[9] This is especially critical for the infantry to be effective. Infantry must train to use combat vehicles and helicopters to enhance their mobility, conserve strength, and arrive at the right place, at the right time, faster than the enemy. Swift and agile movement plus rapidity and intelligent flexibility are the essentials.[10]

The institutional Army must embed *Air*-Mech-Strike doctrine, tactics, techniques and procedures into Soldier, NCO and officer schools and professional development programs. Our Airborne and air Assault schools need additional emphasis to support the *Air*-Mech-Strike requirement for movement and tactical employment. The Reconnaissance efforts at fort Knox and Fort Benning require upgrades to existing programs of instruction. Fortunately, minimum Driver, Gunner, Commander and maintenance training will be required for transformation to the *Gavin*, AGS, *Ridgway* and CSV ATVs. Learning 3-D maneuver skills so they are precursor drills to support unit collective maneuver capabilities would be integrated during and following initial New Equipment Training (NET) by qualifying Soldiers in external sling-loading and internal loading their vehicles with both U.S. Army helicopters and USAF fixed-wing aircraft. These tasks would support unit Mission Essential Task List (METL) requirements re-certified annually. Airborne units would have the additional METL tasks of assisting Riggers in preparing their Air-Mech-Strike vehicles for heavy parachute drop operations.

Commanders must ensure that *Air*-Mech-Strike infantry are trained the same for dismounted tasks. The "One Infantry" concept is critical today, with the Army as small as it is, and the increased demands on infantry Soldiers. There is no room for five different types of infantry, except specialty skills (*Master Gunner, Ranger, Airborne*, etc). All infantry should have the same move, shoot , communicate and survive basic skills.

Air-Mech-Strike requires the current Army Aircrew Training Program to break the paradigm of the "just enough" crewmembers to man and meet minimum proficiency Unit Status Reporting requirements. The total pool of trained aircrews needs to be viewed as a singular body and not allowed to be filtered by non-operational personnel management concerns. Large numbers of aircrew members are diverted to "career survival" positions in lieu of their primary flight duties. Reform this paradox, and the aggregate number of aviators in all three components supports the fielding of at least two crews per airframe. An Army Aviation Training Program that places flight proficiency of all aircrews as its top priority will properly integrate and train aircrews to meet readiness requirements for deployment and sustained combat operations. Aircrew training hours will need to be increased to meet the demands of *Air*-Mech-Strike readiness.

The more fluid, the form of war, the more necessary it is for flexibility to be the main characteristic both of the machinery and the training doctrine.[11] Future combat and Stability and Support Operations will require a force like *Air*-Mech-Strike to adapt to the predicted flexibility of operations. A trained and ready *Air*-Mech-Strike force will meet this future requirement.

Our goal in a trained *Air*-Mech-Strike force is to have the deterrent value or combat effect on our adversaries as Patton's Third Army had on the Germans in WWII. Victor Davis Hanson says; "*The greatest threat...was the [unknown] whereabouts of the feared U.S. Third Army. General Patton is always the main topic of military discussion. Where is he? When will he attack? Where? How? With what? Those are the questions which raced through the head of every German General. The tactics of Patton are doing the unpredictable...*[12]"

---

[1] U.S. Department of Defense, Joint Publication 3-0. *Doctrine for Joint Operations*. (Washington D.C.: Government Printing Office, 1995) pp. 1-6

[2] U.S. Special Operations Command, Publication 1, *Special Operations in Peace and War*, (MacDill Air Force Base, Florida: USSOCOM) pp. 2-30

[3] Statement by German Captain von Bechtolsheim at a Fort Sill, Oaklahoma lecture described in Lieutenant General Arthur S. Collins Junior, *Common Sense Training: a working philosophy for leaders*, (Novato, California: Presidio Press, 1978) p. xi

[4] U.S. Special Operations Command, Publication 1, *Special Operations in Peace and War*, p. 5-3

[5] Sun Tzu, Introduction by Samuel B. Griffith; *The Art of War* (Oxford University Press, 1984); pp. 10-51

[6] Norman Kotker, *Genghis Khan*, (*The Quarterly Journal of Military History*, Volume 4, Number 4, Summer 1992), p. 108

[7] Ardent Du Picq, *Battle Studies*, (Carlisle, Pennsylvania: U.S. Army War College Press, 1983), p. 18

[8] Sun Tzu, *The Art of War,* p. 73

[9] (Colonel) Brigadier General S.L.A. Marshall, *The Soldier's Load and the Mobility of a Nation*, (Arlington, Virginia: Assciation of the U.S. Army Press, 1950), p. 7.

[10] *Infantry in Battle*, (Washington D.C.: The Infantry Journal Incorporated, 1939)

[11] (Colonel) Brigadier General S.L.A. Marshall, *The Soldier's Load and the Mobility of a Nation*, p.115

[12] Victor Davis Hanson, *The Soul of Battle,* (The Free Press, 1999) p. 295. Accounts taken from a captured German Colonel in WWII.

Chapter 11

The U.S. Army's *Air*-Mechanized Future

"*Whosoever desires constant success must change his conduct with the times.*"
—Niccolo Machiavelli, Discorsi 1531

The Importance of Vertical Envelopment and "Air-Mechanized" Operations to Future Combat Operations

We have long understood the value of vertical envelopment. Placing a combined-arms organization on the ground in the enemy rear creates a problem he cannot ignore. It also is a way of attacking relatively less prepared and less protected enemy forces while facing him with direct pressure forces he cannot overlook. This combination of pressures leads to a more rapid and complete enemy collapse. In U.S. Army After Next (AAN) wargames, the value of "*Air*-Mechanized" forces[1] was readily apparent. The dramatically short and decisive AAN campaigns would have become longer, less decisive, and linear without such capabilities.

Implications of the Rapid Rise in Defensive Potential Due to Information-Age Technologies

Future defensives can truly be a "*shield of blows*" as ambush and exploit tactics are used. The potency of fires is growing geometrically as technology improves precision warheads, navigation, sensors, and fire control. Meanwhile mobility and the ability to maneuver forces in general is improving only arithmetically. Future defenses will require fewer and smaller forces in static roles. This allows more forces to play dynamic roles. Defending forces are more easily covered by integrated networks of sensors and long-range precision engagement systems. The Soviet Army in its last days was studying the potential of these and called them "*surveillance strike com-*

plexes." These characteristics and qualities taken together combine to make both area and mobile defenses more potent.

As we begin to understand the implications of the defensive bias of the technologies which are most likely to proliferate first, we have a greater urgency to study solutions for defeating them. A professional and determined enemy will use the potent properties of the defense to his advantage. This means that, just as the firepower superiority of World War I military technology led then to linear warfare and long indecisive campaigns, the technologies of our time may bias ground combat in the same way.

The defensive bias of precision weapons and information will challenge ground offensives. Higher than expected degrees of "*overmatch*" will be required to overcome them. In our time it won't be trenches but broad bands of terrain covered by sensors and precision "*shooters*" integrated and "*networked*" by computers. In a well-prepared defense, these integrated elements will be devilishly difficult to locate and defeat.

Picking one's way through a modern defense just on the ground in 2-Dimensions is time consuming and dangerous. Doing so becomes a linear combined-arms fight supported by deep "*shaping*" fires to cause attrition and suppression in depth. Some call this non-linear combat, but it is not. Deep fires, whatever the source, cause damage, but because these fires cannot be rapidly exploited, their effects are not optimized. The enemy is free to decide how best to cope. These actions in depth, if done well, are supportive but are not decisive. Decision occurs as the line of combined-arms action progresses.

The post-WWI solution was large-scale mobile mounted warfare. During WWII a small percentage of the fighting forces of the major armies were capable of mobile, air/ground vehicle-enhanced warfare. And that percentage was enough to break the deadlock of the trenches. The initial problem for mobile units was to penetrate or envelop the crust of the potent defenses to allow rapid movement through them. The dismounted infantry formations made this possible in many cases. Their objective, then, was to penetrate with a large enough force to threaten operationally significant objectives, and to do it sufficiently swiftly to make enemy reaction in sufficient strength impossible. When this was achieved, campaigns were short and decisive.

The solution for overcoming the defensive bias of future defenses will be to take to the air with some of the offensive combined-arms force and fight the enemy non-linearly. It is not likely that 2-D ground-based operations alone will be able to achieve the mobility required. While a ground-based direct pressure general purpose 2-D force threatens his front, a medium-weight "*Air*-Mechanized" mobile protected 3-D force attacks in depth. Today's light forces will not suffice, because they become immobilized when placed on the ground. Mobility and superb situational awareness will be the key to the survival of these forces and an important source of their potency. Based on the evidence of the Army After next wargames, having "*Air*-Mechanized" forces shortens campaigns dramatically. In fact, campaigns on the ground (absent *Air* -Mechanized forces) would have been linear.

In those wargames, air delivery of large forces into and over enemy held terrain was difficult to accomplish. Just as World War II tankers learned to suppress and avoid anti-tank defenses, future commanders of *Air*-Mechanized forces will have to learn to avoid or suppress complex modern air and ground defenses and pass over or through them quickly with the "*Air*-Mechanized" forces of the future. It will be the next higher commander's responsibility to create the conditions for successful vertical envelopment. This has always been so, but future Commanders can focus the means at their disposal much more effectively, when doing so is important. Delivering a Division-sized force of this nature into enemy-held territory may require the concentrated efforts of the entire theater for a number of hours, but the pay-off will be a shorter campaign.

The AAN wargames also taught a lesson analogous to one learned early in WWII about mobile mounted warfare. In May of 1940, the French employment of armored units, equipped with superior tanks, was less effective than the German because they were employed in smaller tactical formations and scattered in support of the less mobile forces. In the AAN wargames, the most effective employment of these units was in formations large enough to rapidly effect campaign out comes. This is not to say that small tactical level vertical envelopments were not beneficial, they were. They were particularly beneficial when facing an enemy in difficult terrain, getting through defiles and over rivers. These facilitated non-linear fighting at the tactical level. But as WWII leaders soon learned, the big pay-off is to face the enemy with a non-linear *campaign*.

The Potential Evolution of "*Air*-Mechanization"

There are several reasons why "*Air*-Mechanized" forces have not been fully exploited up to now. Only until fairly recently have military fixed-wing aircraft and helicopters have had the capacity to lift a suitably armed and protected vehicle. Improvements to both the UH-60 and CH-47 have likewise allowed extra fuel to range out to a wide variety of conditions which would make landing site prediction difficult. The currently programmed U.S. Army Future Transport Rotorcraft (FTR) could lift one mobile and protected future-fighting vehicle and meet the other conditions. The product of the current joint DARPA and Army Future Ground Combat System (also known as the Future Combat System-FCS) program could meet the need for a future air transportable fighting vehicle or an enhanced version of the 17-ton M8 Armored Gun System of our proposed *Air*-Mech Strike Interim Phalanx Force (see Chapter 8 Organization).

Another challenge to large-scale *Air*-Mechanized operations is how to support them with intelligence, communications, indirect fires, and logistics. This was also a challenge for the designers of the large mobile formations of WWII. The solution then was to combine the new technologies of the time to sustain the Airborne/Armored force sufficiently well and long enough to achieve rapid success. The maturation of radio technology solved the communications problem. The aircraft helped solve the intelligence and fire support problem. And motor transport added logistical viability. And in many cases, Army armored formations themselves were potent enough to achieve rapid success. When rapid success, as in the German campaign into Russia, was denied, problems of support rapidly accumulated.

The *Air*-Mechanized force of the future can draw on the conceptual legacy of its forbears. Satellite and other long-duration overhead sensors can help solve the intelligence problem. Sophisticated overhead platforms will assist in maintaining communications and digital connectivity. Air Force and Army aircraft and longer-range rockets and missiles can provide very focused long-range precision fires and suppression. The logistical solution depends on a combination of advances. First it depends on sufficient planning information to limit the *"just in case"* burden. Then it depends on advances in equipment reliability and consumption frugality to further limit the transportation burden. Finally, it depends on the potency of the force, and the tactical and operational art of those who employ it, to reach a quick decision. (Each of these statements deserves an explanatory paragraph in the longer version.)

Air-Mechanized forces could evolve along two complementary lines. One is delivered by Air Force aircraft (the Advanced Technology Transport -ATT follow-on to the C-130) by parachute airdrop and/or to remote unprepared assault strips and another is delivered by rotary-wing, tilt-rotor, or other vertical take off and landing aircraft (the FTR follow-on to the CH-47). The first is based on a fairly well protected under 20-ton vehicle organized as light armored cavalry for open to mixed terrain missions. The other is based on a lighter vehicle family organized to fight in close and urban terrain. These forces are most successful when task-organized for the specific mission.[2]

An important design criterion for future *Air*-Mechanized forces will be their ability to deploy rapidly across strategic distances. In fact, an important goal should be to make them capable of operational maneuver from strategic distances. In other words, these formations should be able to launch from outside the theater of operations and land in the vicinity of their operational objectives ready to fight.

One of the limitations of the current 101st Airborne Division (Air Assault) is that the bulk of its aircraft cannot fly strategic distances.[3] This limits its strategic utility. Future *Air*-Mechanized formations should be fully deployable through a combination of Corps level Army and theater-level tactical Air Force aviation. If they require intermediate staging, this should be outside the "*operational exclusion zone*" of any potential enemy.

When vertical envelopment by ground operational maneuver-capable forces becomes available, the change in warfare can be as dramatic as that between WWI trench warfare and the WWII introduction of *blitzkrieg*. What could potentially be long and very linear ground campaigns could become very short and non-linear ones. The expense of these formations will be high, but their value has to be seen in a broader strategic perspective.

Road to *Air*-Mechanization

Many factors play into the timeline of a modern military force undergoing a conceptual transformation. The current stress of real world deployments, the perceived threats to national security, budgetary constraints, political concerns from the legislature and the decision-making cycle of the executive leadership. The U.S. Army's role towards *Air*-Mechanization will be no different. The Army itself will need time to assess the mission characteristics of 3-Dimensional maneuver. Analysis will need to be done on equipment compatibility, tactical procedures, organizational reviews, training impacts and future acquisition programs. This book is a good start point intended to modify the current Army transition process underway to lighten the Army force structure to one that facilitates 3-Dimensional maneuver as well. Where do we start? What are the steps and in what order? We propose the following as a possible frame work or road map leading to the force design postulated in numerous Army After Next year 2025 wargames.

Step One: Make the Decision

Before any further serious work can progress toward *Air*-Mechanization the Army needs to decide that a strategically *Air*-Mech capable Army is the end state of the current objective force with full fielding underway beginning around the year 2015. Such a land mark decision will eliminate the counter-productive 2-D only force supporting programs underway and help focus all agencies within the Army on developing supporting efforts to meet the 2-D/3-D end-state goal.

Step Two: Validate all Current Acquisition Programs for *Air*-Mechanized Capability

The move towards *Air*-Mechanization will undoubtedly render several acquisition programs obsolete or in need of significant revision while others may only need small revisions. The limited time available and constrained funding environment underscore the need to prevent wasting time and resources on obsolete 20th century force requirements. One such example would be the current plan to field light armored cars for an interim Brigade test bed at Fort Lewis Washington. Armored vehicles contending for the fielding contract need to be at least sling-loadable by CH-47D/Fs by weighing less than 11 tons. In Chapter 7 we showed how the design inefficiency of wheeled platforms does not justify these vehicles being over the lift capacity of Army CH-47D/F helicopters. A similar approach was used in the early 1960s when the Howze board was formed to evaluate requirements for a helicopter mobile division. Equipment found to be too heavy for the CH-21 to sling-load was eliminated from the organization and lighter substitutions were found. In previous chapters we have proven how tracked M113A3 *Gavins* and *Wiesels* are more capable armored vehicles than the armored cars being considered in all categories at less cost while being capable of Army helicopter air transport.

Step Three: Conduct a Squad-Platoon Level *Air*-Mech Test

This test could be conducted within 90 days of a decision and should be executed in Europe using U.S. helicopters to evaluate the German, British or any other foreign or domestic *Air*-Mech model vehicles. The test would be used to determine which family of vehicles best suits U.S. Army helicopter transportability and ground maneuver requirements. Our book has concluded that the most likely winners of such a competition would be the German *Wiesel* and the United Defense lightweight M113A3 vehicle combinations. Of course our conclusions are based on specification data comparison, our own field experiences and not on formal scientific testing conducted for the purpose of *Air*-Mech-Strike. Let the best vehicle(s) win.

Step Four: Conduct a Series of Company and Battalion *Air*-Mech Tests

This step would be a modification of the current test bed Brigade initiative underway at Fort Lewis Washington. The 1st Brigade 25th Infantry Division would adopt our recommended force structure for a lighter all *Air*-Mech capable unit centered around the lightweight *Ridgway or M113A3 Gavin* series and CSVs. The 1st Brigade 2nd Infantry Division would adopt the 2-D M1/M2 and 3-D M113A3 model we proposed to include the Brigade's RSTA Squadron with the (*Wiesel*) *Ridgway* Fighting Vehicle (RFV). The 1-25th Brigade would be built along the RFV/CSV model for developing the doctrine for converting the 101st Airborne (Air Assault) Division. The M113A3 GFV BN in the 1st Brigade 2nd Infantry Division would model *Para-mech* for the 82d Airborne Division (parachute forced-entry) and working with M1/M2s act as a test bed for the conversion of the heavier Divisions like the 1st Infantry and 3rd Armored. Separate Cavalry Regiments may adopt a combination of the two to embed a degree of flexibility in the total force structure.

Step Five: Develop a Timeline and Capability Milestones

This is perhaps the toughest step of all. Where does the Army begin shifting resources to follow-on equipment and force structures and cease developing the so-called "*legacy*" model? National defense requirements will still have to be met during such a transition further complicating the pace and degree of upgrade. We recommend the following rough outline;

Mixed Range *Air*-Mech Capability 2000-2005 (Interim *Phalanx* Force Phase I)

Air-Mech-Strike strategic reach would be provided by Paramech forces using parachutes or STOL airland operations to deploy GFVs, AGSs, RFVs and CSV s from USAF fixed-wing aircraft which can fly for thousands of miles via mid-air refueling. The lowest common denominator lift aircraft is the C-130H/J Hercules which can airdrop/airland one 34,000 pound M8 AGS 105mm light tank to an unrefuelled 2,000 mile range. However, once in theater, the lift capacity of the current U.S. Army UH-60 and CH-47 helicopter fleet limits the *Air*-Mech vehicles to weights ranging from the *Ridgway* Fighting Vehicle (6,000-8,000 pounds) up to the a lightweight M113A3 *Gavin* Fighting Vehicle (20,000 pounds). The range of these 1960/70 technology helicopters limits the radius of air inserting these lightweight armored vehicles to about 150-200 kilometers via conventional sling-loading techniques. These performance parameters mean that today without USAF fixed-wing aircraft support, U.S. Army *Air*-Mechanized maneuver is capable of relatively short-range tactical operations with light armored vehicles. In addition, they are limited initially to conventional sling-loading (except for RFVs which can be carried internally by CH-47D/Fs) which reduces range and increases risk from enemy air defenses. However, despite these challenges, the advantages of going 3-Dimensional even with lightweight armored vehicles outweighs the risks of being limited to just 2-Dimensions to break through increasingly difficult surveillance-strike-complexes. Displaced Drop, Landing and Assault Zones (DZs, LZs, AZs), small-arms armor protection and mechanized speed once inserted greatly out paces conventional dismounted Airborne/Air Assault techniques with less risk to the aircraft via displaced landing sites. Recognizing the limitations of these light armored vehicles and their short range of insertion by Army helicopters, the prudent application of this first generation *Air*-Mech capability that we promote is to apply the capability across-the-board to all combat formations. Heavier Divisions would gain supporting *Air*-Mech Battalion-size 3-D capable task forces and the lighter Divisions would have all their Battalions ward where the wings pick up a portion of the load that carries the aircraft allowing the rotor to be slowed down which delays the effects of retreating blade stall. The result is the speed increases from a typical helicopter cruise of 130 knots to about 170 –200 knots. This results in a two to three day self-deployment timeline overseas and comes at a significant cost in fuel consumption and maintenance complexity similar to that of a tilt-rotor. The compound helicopter option is probably the lowest-cost FTR but has less potential for mission performance.

The most promising design is a removable and/or retractable rotor configured as a small sized but powerful "*Power sled*" termed the "*Speedcrane*" concept. Several manufacturer's experimented with retractable rotors in the 1960s to include some large-scale wind tunnel testing. The military however, deeply involved in the Vietnam conflict, was more interested in payload than speed and had no requirement for self-deployability. The *Speedcrane* looks similar to a DC-9 with a pair of low fuselage placed swept wings and a pair of turbo fan engines mounted at the rear. The engines are both turbofan and turboshaft with a gearbox designed to shifty power from the outer high bypass fans to a drive shaft to power the main rotor.

The *Speedcrane* may employ a CH-53E transmission, which greatly reduces developmental costs turning a CH-53E based main rotorhead. The rotor system employs a removable or switchblade type rotor blade that folds to half the blade span. The *Speedcrane* carries its payload in pods secured to the under-fuselage in a similar manner to the CH-54 *Skycrane*. Large "*stork*" landing gear provides the ground clearance. On take-off from a base in North America, the *Speedcrane* looks very much like a typical airliner and takes off as a fixed-wing aircraft. The rotor in completely hidden on the back dorsal of the fuselage. The aircraft without the drag of a turning rotor achieves the airliner cruise speed of 450-500 knots making the *Speedcrane* the only FTR contender to have same-day, cross-ocean self-deployability. The *Speedcrane* can fly as a turbofan jet and then land and attach rotors for tactical missions. Or a few minutes out from landing, the *Speedcrane* slows down and dorsal covers open and slide alongside the fuselage exposing the CH-53E rotor system. RPM is built up as power is transferred from the fans to the rotor system. The blades switch blade outward and clutched to power to spin so the aircraft assumes the flight profile of a compound helicopter. Ducted-fan thrust on the rear mounted engines counters the torque of the rotor. Since the rotor only operates part time at take-offs and landings, and even then usually only during tactical missions, the maintenance costs of the *Speedcrane* would probably lower than that of a conventional CH-53E. The fewer parts, fewer drive shafts and a single transmission of a proven design will mean that the *Speedcrane* will likewise have low flyaway costs as well as low developmental costs.

Getting More Strategic Lift: Wing-In-Ground Effect Craft

The strength of the *Air*-Mech-Strike concept is that it puts together the absolute best combined-arms weapons together and doesn't compromise to just one type arm or one type platform. Its an optimized force structure that concentrates on just a few capable platform types and arranges them into packages that are deployable by available strategic and tactical lift. However, its implied that it would be wise to get more strategic lift in order to get the heavy 2-D forces in the *Phalanx* Force to the battlefield as quick as the lighter 3-D forces that are better sized for air transport; expecting these forces to work together in their unique ways later on.

United States military strategy is based on certain explicit assumptions, which influence the environment in which the Army will operate over the next decade[4]. These include a defensive strategy during which the Army will reshape, train and respond to missions. With the end of the Cold War, the threat of a global war receded and debate resumed whether the United States needs to continue to prepare to fight two major theater wars simultaneously. No major peer competitors should emerge over the next two decades; however, the emergence of a coalition of states hostile to the United States could emerge as a threat by the end of the decade. The most probable threats to U. S. national interests that are likely to come will be from failed states, transnational actors and competitors for resources. The United States Army will find itself fulfilling a wide range of missions throughout the globe. The bulk of its forces will be stationed within the United States, but they will deploy on force-projection missions. In this regard, the United States Army has a marked interest in overcoming the "*tyranny of time and distance*" by first organizing itself for 3-Dimensional maneuver on the battlefield with *Air*-Mech-Strike and by improving its deployment mobility means. As USAF General Walter Kross, CINC Transportation Command, pointed out, while aircraft may deploy some forces and their equipment to distant theaters, sealift will continue to be vital since "*95 percent of dry cargo and 99 percent of liquid cargo will likely move by sea.*"[5]

In no other theater is the tyranny of time and distance so evident as that in the Pacific. As Admiral Dennis C. Blair, U.S. Navy Commander in Chief, U.S. Pacific Command, noted during his testimony before the House Armed Services Committee (March 1999), fostering a more secure Asia-Pacific region remains the primary goal of PACOM and that "*deployed, ready, and powerful Pacific Command forces*" are the best foundations for the region's security and development.[6] Pre-positioned stocks and forward deployed forces are the first echelon of American engagement and security in the region, but only linkage to strategic forces in the continental United States can insure the nation's ability to sustain commitments and engage in effective compellence. There is an older *Air*-Mech technology that, while no "*silver bullet*", could assist the United States Army in performing its power projection missions in the Asia-Pacific area and other regions of the globe.

The Tyranny of Time and Distance in the Pacific: One Hundred Years and Continuing

In 1998, the United States Army marked a century of engagement in the Pacific and Far East. In that century, the Army proved a key factor in American forward presence and power projection during peace and war. Its presence was necessary to deter conflicts, work with allies and friendly states, support humanitarian assistance, and win the nation's wars in this vast region. This year, Americans will mark the fiftieth anniversary of the beginning of the Korean War. Korean War historian and U.S. Army combat veteran, T. R. Fehrenbach observed:

> *"Americans in 1950 rediscovered something that since Hiroshima they had forgotten: you may fly over a land forever; you may bomb it, atomize it, pulverize it, and wipe it clean of life, but if you desire to defend it, protect it, and keep it for civilization, you must do this on the ground the way the Roman legions did, by putting your young men into the mud.*[7]

A century of the U. S. Army engagement in Asia began with the Spanish-American War. U.S. naval power might destroy Spain's Pacific Squadron but it could not occupy and hold the Philippines. There was a long delay between Dewey's victory at Manila Bay on 30 April 1898 and the eventual arrival of an U. S. Army force in the Philippine Islands—there certainly were no *Air*-Mech-Strike capabilities available then. This delay created a political-military sovereignty gap and allowed an insurrection to grow which opposed incorporation into the United States. The first of three contingents from Major General Wesley Merritt's Philippines Expedition left San Francisco on 25 May and arrived in Manila on 30 June; the last contingent arrived on 25 July.[8] This delay in the arrival of the expedition allowed Emilio Aguinaldo, a Filipino nationalist, to begin an armed struggle for national independence. This struggle led to a full-fledged insurgency against American rule, which was not completely suppressed until 1902. While the role of the United States and its Army in the Asia-Pacific region has changed over that century, the continuing tyranny of time and distance in the Asian-Pacific area still dominates strategic plans and concepts due to our dependence on sea-lift. By World War II, sailing times had only been slightly reduced, since even today it still requires 21 days to move troops, equipment, and supplies by sea from Oakland to Manila and 16 more to reach the western limits of the PACOM and USARPAC Area Of Responsibility (AOR) in the Indian Ocean. A recent report by Secretary of Defense William Cohen observed, air movement times across the Pacific are measured in hours, but sailing times still reflect *"the tyranny of distance — 19 days from Seattle to Thailand, 18 days from Alaska to Australia, and 10 days from Hawaii to Korea."*[9] Pre-positioning materiel, a Cold-War era solution, arose out of shared threat perceptions and alliance arrangements that developed during that era. They mitigate but do not overcome the tyranny of distance and are dependent upon the continuation of those shared interests at a time of dynamic changes in the Asia-Pacific security environment. The Revolution in Military Affairs—primary an advance in mental not physical advantages—has yet to conquer the tyranny of time and distance for U.S. ground forces that must deploy from the continental United States to the far reaches of PACOM's area of responsibility, an *Air*-Mech solution can if its enhanced to bring over large amounts of vital equipment and supplies.

The United State's engagement in the Asia-Pacific region divides into two epochs and an epilogue. That experience demonstrates how vital U.S. Army presence has been in providing regional stability and protecting American interests. The first epoch was dominated by a rivalry between Japan and the United States. China was weak and divided. Russia was unable to defend her Far East. The epoch began with the Sino-Japanese War of 1894 and ended in 1945 with the Japanese surrender on the battleship *Missouri* in Tokyo Bay.

In that half century, the Navy was the center of gravity of U. S. Pacific military power. U.S. War Plan "Orange" (War with Japan) reflected this geo-strategic calculus. The Army's primary role until Pearl Harbor was to defend the Philippines. The Philippines was far from the continental United States and very close to the Japanese Empire. The American defense ended tragically on Bataan/Corregidor in early 1942. It was the worst defeat of American forces during the entire Second World War, though their gallant stand bought months of time for U.S. forces to mobilize perhaps the critical action that won the war. In modern conflicts we will not have the time to convert industry from "*butter to guns*". It was the United States Navy's inability to reinforce the Philippines, after the disaster at Pearl Harbor which sunk much of the fleet unexpectedly by air strikes, that condemned the American and Filipino defenders to an uneven struggle. There is a distinct possibility with space targeting and the proliferation of guided missiles, mines and stealthy diesel-electric submarines that sea-based resupply of far-flung Army units in the Pacific may be interdicted long enough for an aggressor to conquer his neighbor.

During WWII there was time to rebound and the newly equipped American counter-offensive, using both naval and air power proved the decisive instruments in carrying the war across the Pacific. These forces made possible the deployment of U.S. Army and Marine forces in Airborne and Amphibious advances across the Southwest and Central Pacific Theaters. On the verge of the invasion of the Japanese home islands, President Truman decided to avoid inevitable large-scale casualties and employed atomic weapons to force the Japanese government's surrender. Thereafter, nuclear weapons would be a primary factor in the U.S. military presence in Asia and an ingredient in the management and resolution of Asian security issues.

In 1947, the United States granted the Philippines their independence, after securing a naval and air basing agreement with the elected Philippine government. The U.S. supported another counter-insurgency struggle in the Philippines, this time successfully conducted by the Philippines' government against the *Hukbalahap* communist guerrillas. The second half of the century, and second Pacific epoch, was dominated by the Cold War confrontation between the United States and the Soviet Union. This confrontation took on strategic dimensions in the Pacific with the triumph of communism in China, the detonation of the first Soviet atomic bomb, the signing of the U.S.-Japanese Peace Treaty, and the North Korean invasion (with Soviet support) of South Korea. The Cold War was cold in Europe but hot in Asia. Politically, the broad outlines of U. S. Pacific presence were forged by the end of the Korean War. After air-deployed forces from Japan stabilized the front, reinforcements poured in. One of the central problems in the initial period of the Korean war was the timely deployment time of forces from the continental United States to stabilize the defense and to create a strategic reserve to regain the operational-strategic initiative.[10] Its been said that this "*policy war*" or "*police action*" was "*the wrong war, in the wrong place, at the wrong time*". But the United States Army found itself committed to a full-blown war with an intractable opponent half a world away. A negotiated settlement and not military victory defined the end of the contest and many strategic planners were quite certain that future wars would be won by air power and "*massive retaliation*." There would be a military forward presence on the Korea Peninsula, in the Taiwan Straits, and across Southeast Asia as "*trip-wire*" forces for this "*massive retaliation*". The U.S. deployed a large military infrastructure in Asia, especially in the Philippines and Japan. In 1964, in the context of deteriorating Sino-Soviet relations, the PRC tested its first nuclear weapon. At the same time, the United States assumed the burden of opposing Vietnamese communism following the French defeat there. That commitment, which began as assistance to the South Vietnamese counter-insurgency effort, became America's largest and longest war of the Cold War era. A combination of perpetual air and sealift continued the fight but U.S. forces were always out-numbered, unable to mass the force size required to pacify the South with a defensive strategy nor the *Air*-Mech transportation means to quickly mass enough force to invade and subdue the North had that option been pursued.

American withdrawal from Vietnam and the defeat of the South Vietnamese regime led to a new phase of the Cold War in Asia after 1975. Korea remained stable, thanks to on-the-ground U.S. military presence and the economic transformation of the South. Japan became a truly global economic power. Southeast Asia witnessed a series of economic miracles. The U.S. opened relations with China. In this geopolitical context, the United States' rapprochement with China leveraged the Cold War to the U.S. advantage. Playing the "China card" became a vital part of the East-West confrontation as *detente*' gave way to another round of confrontations. China began a market-driven economic transformation although the Chinese Communist Party maintained its political monopoly on power. In the later 1980s and early 1990s, the Cold War ended in Asia with Soviet disengagement, following their domestic crisis and imperial overreach in places like Afghanistan. The U.S. Army in the Pacific played a crucial role in the final victory in the Cold War by providing a credible military deterrence and presence in Asia, especially in Korea.

It is now the end of the first decade of the post-Cold War era. Many changes in the Pacific security environment have transpired and the direction of those changes raise serious questions about the ability of American land power to deploy in a timely and effective fashion into the theater during the twenty-first century. While the United States still retains the use of a vast forward theater infrastructure in Korea and Japan, the new dynamics in the Pacific and Asia raise the prospect of conflict. Instability in Indonesia and the Australian *Air*-Mech-led multi-national military intervention in East Timor, the explosion of nuclear weapons by India and Pakistan, the recent fighting over Kashmir, China's disputed claims to the Spratley Islands and the growing belligerency of China toward Taiwan point towards the possibility of military conflict in the region. Open discussions of an alliance among Moscow, Beijing, and New Delhi to counter what its architects call "*globalism and U.S. hegemony*" could well be a harbinger of new tensions in Eurasia. These developments makes it imperative that the United States Army develop *Air*-Mech-Strike means to overcome the tyranny of time and distance to maintain credible influence as a projected force in this theater. The United States Army cannot deploy and sustain large forces across the Pacific by ships much faster than it did in 1899. *Air*-Mech means that could expedite movement in the vast Pacific can also expedite deployments from the continental United States to Europe, the Middle East, and the Indian Ocean, by radically shifting the time of deployment over vast distances in crisis situations. Does such an *Air*-Mech prospect exist and can U.S. military strategy and the United States Army benefit from its realization?

Spotlighting an *Air*-Mech-Strike Technological Alternative: WIG

Strategic maneuver is an inherent characteristic of the U.S. Navy and U.S. Air Force. Naval presence has been a feature of sea power since the age of sail. As navies grew to where they could "*command the sea*", they have been able to apply pressure through blockades. Modern naval theory, since Mahan, has viewed the advances of naval technology as enhancing this role. With the end of the Cold War and the decline of its only oceanic contestant for command of the sea, U.S. Chiefs of Naval Operations have championed a new strategic role for the U.S. Navy. This role incorporates precision, deep-strike weapons systems and amphibious capabilities to project power now labeled "*Forward from the Sea*" as an instrument of coastal, littoral warfare. Air power champions since Douhet, Trenchard, and Mitchell have recognized the decisive influence of command of the air and the deep strike capabilities of Air Forces; some like Mitchell and Kenney have also wanted to air-transport Army ground forces to influence the course and conduct of war farther inland than the littorals. From the flight of the experimental B-15 to Latin America on a humanitarian mission in the 1930s to the advent of modern, nuclear-armed, inter-continental bombers and ballistic missiles, strategic aerospace mobility has been a vital component of United States national strategy, as air-deployed U.S. Army forces have helped to rapidly win conflicts in the Dominican Republic, Grenada, Panama, Haiti, Bosnia and now Kosovo. But these Army forces have been "*light* " due to airlift constraints and have required reinforcements arriving very slowly by sea or rail.

The Revolution in Military Affairs and the advent of stealth aircraft and deep, precision-strike conventional weapons has given the United States Air Force the capability to exercise "virtual global presence." B-2 strikes launched from Whitman AFB, Missouri against targets in Yugoslavia were a manifestation of this capability. Both the Navy and the Air Force possess the ability to deploy and sustain credible combat capabilities into distant theaters in a timely fashion. Forward infrastructure provides support and sustainment in many regions of the globe. Each service has its own transport that allows it to deploy worldwide. Naval forces give the U.S. Marine Corps the ability to fight abroad, if nearby or if there is time to set sail to get there.[11] However, the Marine Corps lacks the critical land power mass/expertise with which to engage far inland strategic maneuver in distant theaters. In contrast, the U.S. Army has the critical mass/experience to conduct such maneuver but lacks the strategic mobility to overcome the "*tyranny of time and distance*" to back "*virtual presence*" with actual "*boots on the ground*" presence that changes governments and gets people to stop fighting and killing each other. Until the United States Army acquires the capability to deploy significant land power into theater in a timely fashion, the United States will not have a truly joint force posture to address the full-spectrum of operations confronting the United States in the post-Cold War world.

With the end of the cold war, the Army changed from a forward-deployed power to a primarily CONUS-based force-projection power. The U.S. Army is dependent on the U.S. Navy and U.S. Air Force to get it to the fight on time. Yet, there have been no sweeping concurrent changes in the transport capability of the U.S. Navy or U.S. Air Force to support this new U.S. Army mission. Therefore, despite the best efforts of the sister services, the full weight of the enormous combat power of the U.S. Army is essentially a not brought to bear quickly in a far-off fast-breaking situation. The Army Chief of Staff, General Eric K. Shinseki, has recognized this problem and has moved to address it. During a time of high OPTEMPO, General Shinseki has articulated a vision for the twenty-first century: "*Soldiers on point for the nation transforming this, the most respected Army in the world into a strategically responsive force that is dominant across the full spectrum of operations.*"[12] He addresses the Army's serious logistics problem:

> *Today, 90% of our lift requirement is composed of our logistical tail. We are going to attack that condition both through discipline and a systems approach to equipment design. We are looking for future systems which can be strategically deployed by C-17, but also able to fit a C-130-like profile for tactical intra-theater lift. We will look for log support reductions by seeking common platform/common chassis/standard caliber designs by which to reduce our stockpile of repair*

parts. We will prioritize solutions which optimize smaller, lighter, more lethal, yet more reliable, fuel efficient, more survivable solutions. We will seek technological solutions to our current dilemmas.[13]

In line with this vision, the Chief of Staff ordered the creation of a test-bed medium Brigade which can rapidly deploy on current navy and air force vessels and aircraft. This Brigade, outfitted with new equipment, should reduce logistics tonnage requirements by 50-70% and allow the Brigade to deploy anywhere in the world in 96 hours. Further, the Army should be able to *Air*-Mech a Division composed of these Brigades within 120 hours and five Divisions of these type Brigades within 30 days. This may be possible considering current and future air transport requirements. However, there is the problem of building such "medium weight" Brigades from scratch—a costly effort that the Army projects will amount to about 5 Brigades by the end of the current CSA's tenure. *Air*-Mech-Strike can build medium-weight Brigade Combat Teams much faster and inexpensively than this; transforming the ENTIRE U.S. Army in the same time that only a handful of Brigades can be built from scratch. This would create the forces we need to dominate in peace and in war, but the deployment time of the multi-Divisional force still reflects the *"tyranny of time and distance"* that has dominated the global reach of land power in the twentieth century, if we cannot provide more *"Air"* to move the "Mech". The current sea lift requirement, which calls for 36 Roll-On/Roll-Off ships, do not represent an effective increase in deployment speed and require the possession of an operational arrival port. Nor is there much help from the troubled U.S. Merchant Marine.

A 1991 Rand Study noted the decline of U. S. merchant marine dry-cargo ships from 300 to 200 during the 1980s and projected a decline of in military sealift capacity (i.e., the Ready Reserve Force, Fast Sealift Ships, and Maritime Pre-positioned Ships) to 475,000 tons by the year 2010. The study recommended modernizing sealift and making it "fast", but the technologies explored were hardly revolutionary. For conventional hulled vessels the term "fast" meant an increase from 20 knots to 30+ knots. A Surface-Effects-Ship [SES] option under study used a catamaran hull with an air cushion and had a speed of 55 knots, but this design was judged too technologically risky.[14]

It was the tyranny of distance that drove General Lesley McNair to radically recast the robust but ponderous "square" infantry Divisions of World War I into more mobile, leaner triangular Divisions that deployed globally and won victories in the European and Pacific theaters.[15] Deployability, however, involved costs. In order to give his infantry Divisions offensive punch, McNair pooled assets to increase combat power. In order to sustain global deployability, McNair reduced the weight of armored forces by using light tank destroyers and medium tanks. These proved inferior in armor protection and firepower when they met the German *Panther* and *Tiger* tanks. The trade-off between deployability and combat power was particularly felt in the initial bitter fighting in the *Bocage* of Normandy. General Shinseki, like General McNair during World War II, faces the problem of making a force projection *Air*-Mech Army more deployable, more capable of maneuver, more survivable and more lethal. The challenge of the new medium *Air*-Mech delivered Brigade is to guarantee that it retains crucial combat power, survivability and endurance for decisive maneuver. The most effective way to guarantee that the Brigade's combat power will dominate stability and support operations and, in case of hostilities, will readily prevail is that the Brigade has some heavy forces organic to it and is backed up by its rapid-deployment Division that should follow rapidly in its wake. Both the *Air*-Mech Brigade and Division (with the exception of parachute deployable units) will rely on the use of airfields/runways for their deployment. Their "*Achilles Heel*"— in a crisis is the 30-day delay in the deployment of a Corps to theater. This delay creates incentives for opposing forces to seek to win before the full force can reach the theater and to engage Army forces in terrain that demands manpower and negates high-tech weaponry. In Europe, rail movement greatly facilitated deployment of U.S. Army ground combat power from Western Europe to the Balkan Theater. This indisputable "ace" in the success of IFOR depended on the staging area in Hungary. However, in many theaters, sealift is still the only way to get large Mech forces into theater. This was true during the Gulf War and would certainly be true of any conflict in the Pacific or beyond. This "way" is not guaranteed to even work if interdicted by modern anti-ship weaponry aided by readily available satellite imagery and its certain to arrive no faster than they did at the turn of the Century; in our *Icarean* world certainly too late.

While the Army experiments with the creation of a lighter, more agile *Air*-Mech force, there is a comparatively old technology that could solve the Army's dilemma by providing rapid, inexpensive, long-range, heavy- lift capability, which does not require a seaport or an airport in which to take-off or land. This technology can transport *Air*-Mech-Strike lightened versions of the Army's lethal Heavy Divisions and their logistics so there is no loss of combat power. This technology can transport this potent *Air*-Mech-Strike force. That technology is Wing-In-Ground (WIG), a proven technology that has been around for 65 years. The Soviet Union experimented with this technology and built a series of *Ekranoplans* [screen glider] for a wide range of missions. Russia continues to support the development of the Ekranoplan for its own navy, other services, and for foreign sales.[16]

Getting There First, with the Most, on the Cheap

Do you want it there fast or do you want it there cheap? This has always been a concern of manufacturers, merchants and logisticians. When the shipment is trans-oceanic, mile for mile, sea travel is the cheapest. Air shipment is faster, but costs five times more per kilogram of weight.[17] However, WIG technology can deliver large amounts of cargo with significantly less fuel consumption (50% more payload with 35% less fuel consumption than similar-sized aircraft 75% less fuel than comparable-sized hydrofoil ferries). Further, the infrastructure requirements for WIG technology is substantially lower than for aircraft or ships.[18] WIG craft travel nearly as fast as aircraft using much less fuel, they are normally based on a body of water, but can take off and land on ground or water and they do not need a developed airfield or port to function.

The Wing-In-Ground Effect refers to the dense cushion of

air that develops between a wing and the water (or ground) surface when they are close together (at low altitude). Sea birds use the WIG effect to skim the water's surface, barely flapping their wings, for hours at a time. Every aircraft experiences the WIG effect as it takes off and lands. Pilots of damaged aircraft conserve energy or use the power of remaining engines more efficiently by dropping down to sea-skimming level to use the WIG effect—although most aircraft are not designed for long-range low-altitude flight. The closer the wing is to the ground (or water), the greater is the amount of lift (and consequently the amount of drag decreases). The larger the WIG *Air*-Mech craft, the more efficient it is when compared with a smaller craft flying at the same altitude. The Wing-in-Ground effect produces the effect of a much larger wing area without actually increasing wing size.[19]

WIG technology has particular appeal to the military logistician. WIG *Air*-Mech craft can move heavy loads rapidly across the ocean and land on an undeveloped beach or further inland. It can fly around bad weather. Since it is flying 3-90 feet above the ocean surface, it is hard to detect using radar, infrared or satellite. It can presently fly in excess of **400 miles per hour and carry over 500 short tons.**[20] It can fly over water, sand, snow or prairie. It can also fly up to altitude of 3,000 meters, but then it loses its fuel-saving advantages. Russian analysts consider that WIG technology is now at the point where the U.S. can build an ocean-skimming WIG *Air*-Mech craft. It would weigh 5,000 tons and carry a cargo of 1,500 tons for a distance of 20,000 kilometers (12,420 miles) at a speed of 400 kilometers per hour (250 miles per hour). Such a craft could deliver 1,200 tons of military equipment and cargo plus 2,000 Soldiers.[21] Russian analysts feel that, with financial backing, they could build a 5000-ton craft capable of lifting 1200 tons or 3000 passengers now. It could fly at 800 kilometers per hour (500 miles per hour) with a range of 16,000 kilometers (9936 miles).[22]

WIG *Air*-Mech craft externally resemble airplanes. They have two huge wings mounted on the hull. The craft uses a turbofan/turboprop or a jet aircraft engine for propulsion. It employs a vertical rudder, horizontal rudder, wing flaps, and a stabilizer to control the craft's heading, and to maintain its flight altitude. Its fuselage and wing structure share aircraft characteristics. Most of it's on board equipment and instruments come from aircraft. Yet, a WIG *Air*-Mech craft is not an aircraft. An aircraft relies on the flow of air past the wings for the lift needed to fly. A WIG *Air*-Mech craft uses the ground effect to fly at a low altitude of between 0.8 and 30 meters above the surface of the sea. Most aircraft cannot do this for extended periods of time.[23]

A Bit of History

Research on wing-ground-effect began in the 1920s. The first WIG craft were patented in Finland in 1935. T. Kaario, the Finnish engineer built what he called the "wing-ram" craft in that year.[24] The Soviets began building such craft in the late 1950s and gave the prototypes the designation Ekranoplan. In 1963, the "*Caspian Sea Monster*" appeared on the waters of the Soviet Union, (See Figure 2). It was 92 meters (100m yards) long with a 37-meter (40.5 yards) wing span and 22 meters (24 yards) high. Nicknamed the *Korabel Maket* [[ship model], it could lift off at 544 tons and cruise at 280 miles per hour. Thirteen 98kN [kiloNewton] turbo-jet engines provided lift and thrust. Eleven of the engines lifted the craft from the water and two provided its cruise power. It took off and landed on water and flew at ten feet above the surface.[25] Due to its shallow draft, it could load and unload in shallow, undeveloped ports.[26] This craft crashed in 1980.[27]

The Soviets went on to build other smaller WIG craft. The first of a planned 120 *Orlyonok* [Eaglet] appeared in 1972. It was 58 meters (63.5 yards) long, had a wing-span of 31.5 meters (34.5 yards) and a height of 16 meters (17.5 yards). It could lift off at 140 tons and carry 20 tons of cargo. Two 98kN turbofan engines provide the lift while an 11.3 MW turbo-prop engine provided the cruising speed of 217 miles per hour at 6 feet above the water's surface. Three of these craft were actually built.[28] The Central Hydrofoil Design Bureau named after R. E. Alekseev, located in Gorky [now Nizhni Novgorod] designed and built the "*Lun*" [Harrier] and "*Spasatel*" [Rescuer] WIG craft for the Soviet and Russian Navy. They also built the small *Strizh* [Martin] WIG trainer craft. At least five other variants of WIG craft were also built. Many of them are still operating safely over the busy waters of the Caspian Sea.

Since the collapse of the Soviet Union, Russia has continued to research, design and produce WIG craft for domestic and international sales. In addition, Great Britain, China, Germany, Finland, Japan, South Korea, Australia, and Montenegro have all conducted WIG craft research and production. The U.S. Air Force considered WIG technology in the 1970s, but built the C-5 that requires runways instead.

China, a great power in the Pacific, is particularly interested in WIG *Air*-Mech technology. Chinese analysts attribute the following advantages to WIG craft over conventional ships and aircraft:

—Superb Mobility. A WIG *Air*-Mech craft travels above the water's surface to travel in the air whose density is 800 times less than that of water. This greatly decreases the drag exerted on ordinary vessels and greatly increases its speed. Fast transports have a top speed of 20 knots. A conventional warship has a maximum speed of 30 to 40 knots, and although the hulls of hydrofoil craft and hovercraft travel above the water, their hydrofoils and their aprons still come in contact with the water. Thus their speed is limited to between 70 and 80 knots or less. But a WIG *Air*-Mech craft can travel between 300 and 400 knots.

—Superb Airworthiness. A WIG craft can fly around bad weather or fly above a stormy sea. Since a WIG craft is not pounded by the storm waves it is remarkably seaworthy. It is also very airworthy.

—Ease of Operation. A WIG craft is controlled through its vertical rudder, its elevator, and its wing flaps. It is simpler to fly than an airplane, and it turns easily. The WIG craft's speed and altitude are easily controlled by the flaps.

—Economical operation. Pressure under the wings of a WIG craft increases greatly by flying fairly close to the water surface. Consequently, only 80 to 130 horsepower are required to propel each ton of weight. The large lift-drag ratio means that fuel-consumption is less and the cruising radius is expanded when compared to similar-sized aircraft. WIG craft are far superior to ordinary aircraft and helicopters in carrying capacity, speed, and cruising radius when using the same power.

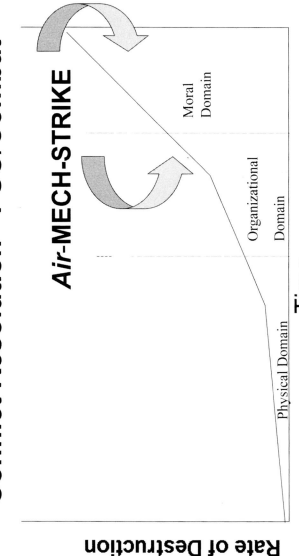

Conflict Resolution SO/Combat with *Air*-Mech-Strike
(General David Grange)

—Convenient Maintenance. WIG craft do not need permanent shore bases. Unlike other high-speed craft, they are able to come ashore under their own power and do not need cranes or chutes. Furthermore, since they have no aprons, like hovercraft, maintenance is very convenient. WIG craft do not have to make a gliding takeoff from the water or land on the water like seaplanes. This lessens the corrosive effect of seawater on the hull.

—Diverse Flight Modes. Not only can WIG craft fly quickly and steadily above water, under radar detection but they can also fly above beaches, marshes, grasslands, deserts, glaciers, and snow-covered land.

—Flight Safety. Should the engines fail, the WIG craft can travel on the water like a conventional ship. They are stable craft, which have operated safely over the years. Some WIG craft vent their engine exhaust forward beneath the wings of the craft to create an increase in dynamic lift. This not only assists takeoff and improves amphibious performance, but also improves flight safety.

—Military applications. The speed, maneuverability, amphibious capability, and reduced signature of WIG craft are greater than that of other craft. Their fast, low-altitude approach may allow them to become the next generation of fast attack craft replacing hydroplanes and hydrofoils. Since WIG craft usually fly within 50 meters of the surface, they are in the radar sweep and search blind zone. The ultra-low altitude of WIG craft leaves no traces on the water surface and is difficult to detect by radar. WIG craft are not optically trackable from space like conventional surface ships. This greatly increases the concealment and surprise attack capabilities of the craft. This extraordinary concealment capability has extremely important military significance. WIG craft may be used as *Air*-Mech landing craft and for the rapid and effective movement of Heavy AFVs, *Gavins/Ridgway* Fighting Vehicles and troops in a campaign. The low flying altitude, the long cruising radius, and the AFV carrying-capacity of WIG craft are second to only ships. WIG craft are also suited for anti-submarine patrol craft, high-speed minelayers, minesweepers, and rescue craft.[29]

Neither Fish nor Fowl

A U.S. Army separate mechanized Heavy Brigade, with all its personnel and equipment, weighs in at 26,649 short tons (69,623 metric tons) and requires 97 x 20-foot containers for conventional shipment.[30] This Brigade could be moved on just 11 WIG craft, each designed to move 2,500 tons. An *Air*-Mech-Strike Brigade will weigh less, but the helicopters take space, so the same number of WIG craft might be needed to move the *Air*-Mech-Strike Brigade. So, why does the U.S. Armed Forces not have WIG craft to move its Army rapidly where it is needed?

The first issue: is WIG a naval craft or an aircraft? The navy has not included WIG craft in its future procurement program, probably because there isn't a surface vessel in the entire navy that can keep up with it. While the Navy did have a long relationship with American seaplane designers from Glenn Curtis to Howard Hughes, the Navy lost interest in seaplane development in the 1950s when it discounted jet-powered seaplanes as a nuclear bomber platform. Interest in transport seaplanes ended a decade earlier with the abandonment of Howard Hughes H-4 *Hercules* prototype— a project designed to enhance strategic deployment capabilities over long distances. There was one flight by Hughes enormous "*Spruce Goose*" flying boat. On November 2, 1947 it flew at 70 feet over the water for a distance of one mile at a top speed of 80 miles per hour. It was the first and only example of a large-platform WIG mission in U. S. history.[31] Successful WIG development could pose a serious challenge to existing naval platforms because WIG warships would have tactical, technical characteristics far superior to existing surface warship classes and a naval race over the application of WIG technology to warfare at sea could negate capital advantages that the United States Navy enjoys with its current surface combatants.

The Air Force is also not interested and does not procure transport aircraft that can routinely operate off of dirt or water. The Air Force prefers to operate only off of permanent hard stand airfields. The need for rapid strategic deployability, which drove the development of Hughes flying boat, is once again a chief concern for U. S. defense planners and a major consideration in the ability to make the United States Army into a full-spectrum, power projection force.

Since the WIG craft do not fit neatly in either the Navy's or Air Force's "comfort zone" and since the Army is the only service without strategic mobility, perhaps the WIG craft belongs in the Army as part of Army aviation or transportation corps. With WIG craft, the Army could *Air*-Mech its heavy AFV elements rapidly to the crisis area regardless of the presence or lack of secure ports and airfields. The Army could deploy with full combat power while the Navy and Air Force could continue their traditional Title X roles by providing longer-term logistics support. WIG technology is not new and other countries are adopting it. Who operates the WIG craft is not important—what is that we obtain them to transport Army *Air*-Mech-Strike forces. It is time for the United States to embrace this WIG *Air*-Mech technology and provide strategic mobility to its Army.

WIG Can Be a Deployment Enhancement in a Time of Need

Returning to the issue of overcoming the "*tyranny of time and distance*" in the Pacific and other theaters, it is very important to note that WIG technology should not be treated as a "silver bullet". But it does represent a potential force-deployment enhancement at a time when the United States retains a wide range of distant commitments and faces the prospect of serious declines in forward infrastructure. Upcoming negotiations with Japan over defense burden- sharing may provide some indications on the probable scope and scale of U.S. defense infrastructure that will be in place at the end of this decade.[32] In South Korea the government has undertaken an expanded defense burden under the assumption that it must prepare for the inevitable withdrawal of U.S. forces from Korea.[33] This comes at a time when the North Korean regime seems to have stabilized its domestic situation and continues to pour resources into its military establishment a point made by General Thomas A. Schwartz, CinC United Nation Command/Combined Forces Command and Commander U.S. Forces Korea, in his recent testimony before the Senate Armed Services Committee.³⁴ Recent increases in the People's Republic of China's Defense Budget and greater stridency over the issue of Taiwanese independence have gone hand-in-hand with a developing arms race in Asia and point to prospects for conflicts in the region.³⁵ When the destabilizing developments in the Indian Ocean are added, the requirement for the United States Army to overcome the "*tyranny of time and distance*" and be the cornerstone of U.S. power projection in the Pacific becomes apparent. There are compelling reasons for a second century of American presence in the Pacific and elsewhere. WIG might just *Air*-Mech us there.

---

¹ "*Air*-Mechanized" or air-delivered mobile protected fighting units were thought of as early as the 1930's by the leading military reformers of that time. Soviet Marshall Tukachevski wrote about them as an extension of the Soviet version of *blitzkrieg*. The British, German and Russian armies today have a such units. But the technologies to fully exploit such organizations of this nature are just now emerging.

² While it is too early to advance organizational formulas at this time, the AAN Tactical wargaming during the summer and fall of 1999 by TRAC at Ft. Leavenworth revealed some important insights about organizational requirements and tactics for these type forces.

The minimum required range for tactical aircraft to self deploy is thought to be about 2100 nautical miles. This allows them to "*island hop*" or makes air-refueling practical.

³ The minimum required range for tactical aircraft to self deploy is thought to be about 2100 nautical miles. This allows them to "island hop" or makes air refueling practical.

⁴ A version of this section on WIG appeared in the July-August 2000 U.S. Army *Military Review*.

⁵ Walter Kross, The Joint Force Commander and Global Mobility, *Joint Forces Quarterly*, (Spring 1998), p. 61.

⁶ Written Statement of Admiral Dennis C. Blair, USN Commander in Chief, U.S. Pacific Command, HOUSE ARMED SERVICES COMMITTEE ON FISCAL YEAR 2000 POSTURE STATEMENT, 3 March 1999, *http://www.pacom.mil/homepage.asp*

[7] Harry G. Summers, Jr., The Korean War: A Fresh Perspective, *Journal of Military History* (April 1996), http://www.thehistorynet.com/MilitaryHistory/articles/0496_text.htm, citing: T. R. Fehrenbach, *This Kind of War*, (New York: Macmillan Company, 1963), pp. 659-660. The U. S. Army Pacific, charged with the ground mission within U.S. Pacific Command has made this Fehrenbach's quote its masthead statement at *http://www.usarpac.army.mil* Fehrenbach invoked the idea of a professional Army acting as a constabulary force as a necessary attribute of U. S. national power during the Cold War. Current operations tempo and strategic engagement make that point even more apparent in this post-Cold War world.

[8] Brian McAllister Linn, *The Philippine War, 1899-1902* (Lawrence: University Press of Kansas, 2000), pp. 13-15.

[9] U. S. Department of Defense, *The United States Security Strategy for the East-Asia-Pacific Region* (Washington, DC, 1998), p. 11.

[10] James F. Schnabel, *Policy and Direction: The First Year. In: United States Army in the Korean War* (Washington, DC: United States Army, Office of the Chief of Military History, 1972), pp. 80-99.

[11] Ibid..

[12] Eric K. Shinseki, General, Chief of Staff of the Army, Address to the Eisenhower Luncheon, 48th Annual Meeting of the Association of the United States Army, (12 October 1999), http://www.hqda.army.mil/ocsa/991012.html.

[13] Ibid..

[14] Myron Hura and Richard Robinson, *Fast Sealift and Maritime Prepositioning Options for Improving Sealift Capabilities*. (Santa Monica: Rand, 1991), pp. iv-xi. Short-term advantage in initial speed of SES deployment are quickly overcome by the improved conventional transports cargo tonnage over time. The marginal speed advantage of SES model is clearly not worth the costs/risks. But a SES is not wing-in-ground, which offers a radical increase in speed and enhanced cargo handling capabilities. These WIG vessels would be designed to meet military cargo requirements in order to enhance speed of loading and unloading and would be designed to work over the beach and not through congested ports.

[15] Russell F. Weigley, *History of the United States Army* (New York: MacMillan, 1967), pp. 461-471.

[16] V. I. Denisov, Spasatel Search and Rescue Ekranoplan, Sudostroenie, (January 1995), pp. 9-12. FBIS: FTS19950101000928; Viktor Mikhailovich Ratushin, 17 August—Air Force Day: The Border Knows No Calm, Just as the Border Troops Aviation Does Not, Armeiskiy sbornik, (August 1997), p. 27. FBIS: FTS19971031000428; and "Opening statement" by E. Primakov, Minister of Foreign Affairs of Russia at the Fourth ASEAN Regional Forum Meeting in Kuala Lumpur 27 July 1997, FBIS: FTS19970727000502. The Spasatel search and rescue ekranoplan has a displacement of 500 tons and has been developed for the Russian Navy to provide rapid search and rescue capabilities in the case of major maritime disasters like the loss of the SSN Komsomolets in 1989. Russia's Federal Border Guard Service, according to its commander General Lieutenant Ratushin is studying the development of both ekranolet and ekranoplan technology. Foreign Minister Primakov used the ASEAN summit in 1997 to support the development of a regional search and rescue center employing the Russian-made "Spasatel" system. The Spasatel is a conversion of the Lun combat ekranoplan, which was designed by the R. Ye. Alekseyev Central Design Bureau of Hydrofoils Scientific Production Association (now the Central Design Bureau of Hydrofoils Joint Stock Company and the Volga Shipbuilding Factory Joint Stock Company). According to its technical specifications, Spasatel has a weight of 400 tons and a cruising speed of 300-400 km/hr. It is powered by eight two-stage NK-87 turboprop engines, and can carry up to 500 persons and has a range of 3,000 km.

[17] Radacraft, Why are Ground Effect Craft being developed? *http://home.mira.nte/~radacorp/index.html.*

[18] William J. Greene, *The Imminent Future of Ultra-Fast Ferries is Off the Water: Breakthrough Design Offers Better Efficiency & Maneuverability*, Paper presented to the International Conference on Air Cushion Vehicles and Wing In Ground Effect Craft sponsored the Royal Institution of Naval Architects, London, 3-5 December 1997.

[19] Ibid.

[20] Ibid.

[21] M. Boytsov, The 21st Century and the U.S. Navy, Morskoy sbornik [Naval digest], July 1995, 74-78.

[22] Andrei Deryaga, Ekranoplany korabli XXI veka [WIG craftships of the 21st Century], Orientip [Orienteer], December 1999, 64.

[23] Zhang Xu, *The Development and Future of Wing-in-Ground Craft*, Jianchuan Zhishi [Naval and merchant ships], 8 September 19 95, pp 30-31. [FBIS translation].

[24] Greene; and L. S. Shapiro, Samye bystrye korabli (Leningrad: Sudostroenie, 1981), pp. 147-151. As was Soviet practice to the late Cold War, Shapiro discusses ekranoplan development in terms of Western prototypes without reference to the work of Soviet design bureaus and shipyards.

[25] Thomas Buro, Ekranoplans-*The Caspian Sea Monster*, http://www.stud.uni-wuppertal.de/~ua0273/ekranoplan.html, extracted from *Ekranoplans—The fast craft of the future?* Schiff & Hafen 9/94, 148-151.

[26] Trottman.

[27] Buro.

[28] Ibid..

[29] Ibid..

[30] Military Traffic Management Command Transportation Engineering Agency, *MTMCTEA Reference 97-700-5, Deployment Planning Guide: Transportation Assets Required for Deployment,* Newport News: MTMCTEA, July 1997, A-8.

[31] The Hughes Flying Boat, *http://sprucegoose.org/sprucegoose/history/history1.htm.*

[32] Joseph Coleman, *Tokyo Argues U.S. Forces Too Costly*, *Washington Times*, (10 March 2000), p. 15.

[33] *South Korea Preparing For U.S. Troop Withdrawal*, (9 March 2000), *http://www.stratfor.com/asia/commentary/0003090043.htm*

[34] Statement of General Thomas A. Schwartz, Commander-in-Chief United Nations Command. Combined Forces command and Commander United States Forces Korea, before the Senate Armed Services Committee, 7 March 2000

[35] Charles Hutzler, *Chinese Military Gets Large Budget Boost, Washington Times*, (7 March 2000), p. 1; and Harvey

Sicherman, *China's Three Ifs*, Foreign Policy Research Institute [fpri@fpri.org], (4 March 2000); Robert Wall, Space, *New Technologies Shifting Air Force Strategies*, *Aviation Week & Space Technology*, (6 March 2000), p. 51; and Corwin Vandermark, *Chinese Navy One Of Many Arming Up In Asia*, *CDI Weekly Defense Monitor*, (10 March 2000), *cvanderm@cdi.org*.

Chapter 12

Conclusions

"We are not more numerous, but we shall beat you because of our planning; we shall have greater numbers at the decisive point. By our character, our energy, our knowledge, our use of weapons, we shall succeed in raising our morale and in breaking down yours"[1]
—Marshall Foch

The U.S. Army is required to be a strategically responsive force across the full-spectrum of operations. Our National Military Strategy requires strategic agility, overseas presence, power projection and decisive force. These requirements are necessary to conduct Smaller Scale Contingencies (SSC) and Major Theater Wars (MTWs). Currently our forces are either light or heavy; the light forces needing increased lethality, survivability and mobility; and the heavy forces needing increased deployability and agility. The current approach is to create separate "medium" type Brigades that are heavier than light and lighter than heavy. This solution improves strategic deployability, increases the firepower for the light force, enhances protection and in most situations can operate satisfactorily across a wide range of scenarios. What it does not do is ensure that there is a strategic and tactical capability to put forces in the right position to fight, attaining "*positional advantage*" over the enemy. This can only be achieved if our forces have a 3-Dimensional, combined-arms capability that provides the commander with the ability to move faster, cross obstacles and exploit the enhancements of situational awareness; in order to take full advantage of the enemy's disposition. We need to have this capability across the entire force structure with multi-functional Divisions, Brigades and Battalions to meet the increasing optempo of a small Army.

The current state of mobility, maneuver options, firepower,

survivability and lethality has not changed much since the Korean War. Capabilities are relatively the same for light units with the exception of technology enhancements including better weapons, communications and a further reach for air insertion of dismounted troops via the helicopter. Mobility, increased staying power and survivability have only seen moderate improvements. The survivability and lethality of heavy forces has increased tremendously, but at the same time, the mobility and versatility of these forces has in fact *decreased* as these forces continued to get heavier. When U.S. Forces were contemplating how to attack through mountains and heavily mined roads and defiles in Kosovo, the lack of a 3-D force made the options fairly limited. A capability to gain positional advantage was not possible

For Airborne operations we have improved Paratrooper lethality, air insertion capabilities. and the technological superiority of U.S. Air Force aircraft. We have not improved protection (survivability of our forces once on the ground), mobility once inserted nor mounted firepower. With the finest helicopter fleet in the world we have not improved internal load capabilities, vehicle and helicopter design to enhance compatibility, nor maneuver doctrine to take advantages of 3-Dimensional operations.

Force planning in this era of uncertainty, requires a set of capabilities to handle the full spectrum of requirements as directed by our National Security Strategy. Elements of both threat-based and capabilities-based force planning is necessary during these periods of uncertainty.[2]

The principles of war laid out in Joint Pub 3.0 encourage the development and use of the *Air*-Mech-Strike concept:

Objective — *Air*-Mech-Strike is focused on destroying the enemy's capability and will to resist.

Offensive — *Air*-Mech-Strike provides additional means to maintain the initiative and freedom of action.

Mass — *Air*-Mech-Strike allows massing affects throughout the depth of the battlefield.

Economy of Force — *Air*-Mech-Strike is a maneuver means to obtain positional advantage to deliver additional fires, and pose new problems on the enemy.

Unity of Command — The *Air*-Mech-Strike force organization provides commanders at Division, Brigade and Battalion level the capabilities, local training advantages and common doctrine to prepare and fight as a Combined-Arms Team.

Security — *Air*-Mech-Strike provides additional survivability (armored protection, additional mobile firepower, and speed of movement) and with enhanced reconnaissance capabilities, the ability to gather critical information in depth throughout the battlefield.

Surprise — *Air*-Mech-Strike provides additional means to achieve surprise through the indirect approach, and moving forces to unexpected points quicker then the enemy can predict or react.

The American Soldier is more capable of carrying out his mission and more likely to survive because he has flexible, 3-D capabilities. The "*vulnerability*" myth that has haunted Army Aviation for years is being confused with "*survivability*". Vulnerability is only an input to survivability. There is no doubt that 3-D maneuver is vulnerability, but those vulnerabilities can be overcome when properly enhanced with fire support, proper intelligence, detailed planning and proven tactics and techniques.[3] There is always risk in any operation. The indirect approach optimized by the AMS concept outweighs the survivability concerns of 3-D maneuver with decisive victory.

Air-Mech-Strike provides a maneuver force the means to attack decisive points that are key to protected enemy centers of gravity, otherwise out of operational reach. Operations conducted deep into enemy territory, with *Air*-Mech-Strike forces, can also create the impression that there are too many friendly forces for the enemy to handle effectively — a special operations tactic that can be effectively used by our conventional forces[4].

Interdiction options afforded by *Air*-Mech-Strike, complement maneuver, and can either disrupt, dislocate, delay or destroy the enemy with asymmetric actions. Conflict resolution, either during combat, or during Stability And Support Operations, is balanced between time and rate of destruction. Key to operations is the rate of destruction of the enemy physically or psychologically, and the protection from that destruction of your own forces. Napoleon considered the moral (psychological) to the physical in importance, "*as three is to one*". In the history of war, the "moral" form the more constant factors, changing only in degree, whereas the physical factors are different in almost every war and every military situation.[5] J.F.C. Fuller advocated that armies should always maneuver against the enemy's will, especially against the enemy commander — because action without will lost coordination, that without a directing brain, an Army is reduced to a mob.

Additionally, Fuller writes; "*it was that while Alexander's Phalanx held the Persian main body in a cinch, he and his companion cavalry struck at the enemy's will, concentrated as it was in the person of Darius. Once this will was paralyzed, the body became inarticulate.*"[6] *Air*-Mech-Strike is a means to focus on the moral domain (will of the force) of the enemy as well as his organizational domain (command and control, logistical lifeline, communication means, combined arms, intelligence gathering apparatus). To focus our maneuver and interdiction against the moral and organizational domains by shattering enemy morale and physical cohesion; and their ability to fight as an effective, coordinated whole, is more effective than head-on attrition warfare against the physical domain.

It is more effective in time and numbers of casualties to attack the moral domain. In examination of decisive battles in history, in almost all, the victor had his opponent at a psychological disadvantage before the clash took place.[7] *Air*-Mech-Strike dislocates the enemy's psychological and organizational balance, setting conditions for our 2-D heavy force component to complete destruction.

The accelerating destruction of the enemy by attacking the moral and organizational domain is supported by James J. Schneider, who describes this phenomenon as "*Cybershock*".[8] *Cybershock* is a pattern of warfare that causes paralysis by attacking the enemy's nervous system, it's cybernetics. *Cybershock* combined with attrition drives an organized system into disorganization — destroying the coherence, connection and flow of information among the component parts of a complex system. To defeat the enemy rapidly with minimum friendly casualties, *Air*-Mech-Strike proves an optimum maneuver and interdiction means for combat and Stability And Support Operations.

The robust reconnaissance assets of *Air*-Mech-Strike allow obtaining accurate, timely information about the enemy's activities and resources, the environment, and objectives as they relate to the physical, organizational and moral domain in conflict. With detailed analysis, intelligence provided by a robust means of reconnaissance, and the ability to act faster than the enemy with the employment of forces at the right place, at the right time, in the right force mix allows our forces to keep the enemy off-balance. We will take the first moves obtaining positional advantage, and combined with simultaneity of actions will shatter our adversary's morale and cohesion.

Air-Mech-Strike forces with a variety of indirect approaches can dislocate, preempt, degrade, disrupt and destroy enemy forces with a tempo that presents multiple problems to the enemy. When synchronized with our heavy 2-D actions of forces will overwhelm any enemy force.

Some specialization is required in our Army. Special Forces, Rangers and Airborne units are an example. But the majority of the Army should be multi-functional. The Army is too small, with too much to do, to have the variety of specialization that we have today. For example, the two Divisions in Germany would be fortunate to have their organizations shaped like the 2d Infantry Division in Korea. If you add *Air*-Mech to that Division design you have a very flexible, multi-functional unit that can task organize for any environment, enemy and type of terrain. With *Air*-Mech-Strike, you have the ability to form combined-arms teams at Division, Brigade, and Battalion level that provide the capability to strike at the organizational and moral domain of the enemy, not only the physical domain.

If we do as the Russian Army has done – get the weight off the back of the combat Soldier, and put it on some form of all-terrain, air-transportable ground vehicles, we increase both 2 and 3-Dimensional mobility.[9]

Adopting *Air*-Mech-Strike provides a capability throughout the Army with:

- An interim and objective 3-D maneuver force that gives the Army a capability now.
- Enhanced strategic / operational deployability and terrain agility for light and heavy units.
- Increased maneuverability that enhances firepower.
- Increased survivability of the light force.
- Increased staying power of the light force throughout the depth of the battlefield.
- Increased RSTA capability for improved situational awareness.
- Geographic, multi-functional capabilities allowing for increased use of intra-theater resources.

Air-Mech-Strike is Cost Effective:

- Costs less than 2-D options via extensive use of Commercial Aircraft and low acquisition costs.
- Leverages existing combat vehicles with modest technology insertions.
- Takes advantage of the largest helicopter fleet in the world.
- Increased opportunity for combined-arms training locally.
- Selective Battalion reorganization within the Army's AC and RC units.
- Units convert at a rapid pace.

We believe the *Air*-Mech-Strike 3-Dimensional *Phalanx* is a concept that gives our Army a required capability now and into the future. It certainly bridges the gap until the Army After Next changes come to being. *Air*-Mech-Strike focuses on the maneuver, the indirect approach, and allows our forces to attack the enemy's moral dimension with flexible, adaptable options, and minimum loss of life. Military theorist von Clausewitz concluded in his treatise, *On War*, Volume i, page 177:

> *The moral forces are amongst the most important subjects of war. They form the spirit, which permeates the whole being of war. These forces foster themselves soonest and with the greatest affinity on the will which puts in motion and guides the whole mass of powers, uniting with it, as it were, in one stream, because this is the moral force itself".*

As our NATO *Air*-Mech capable allies already know, the supporting technology is yet to be fully developed. As a remedy, we recommend a U.S. Army *"Howze II' Air*-Mech Board be convened to begin developing and testing solutions in coordination with industry and unit commanders. To divert resources towards another 2-D force model will only prolong the current World War II and Vietnam model of mechanized and dismounted Airborne/Air Assault maneuver warfare; at our Army's peril in the face of 21st Century combat eventuality.

---

[1] (Colonel) Brigadier General S.L.A. Marshall, *The Soldier's Load and the Mobility of a Nation*, (Arlington, Virginia: Association of the U.S. Army Press, 1950), p. 112.

[2] John Troxell, *Force Planning in an Era of Uncertainty* (Carlisle, Pennsylvania: Strategic Studies Institute, 1997) pp. 24-39

[3] Lieutenant General John J. Tolson, *Vietnam Studies-Air-Mobility 1961-1971*, Washington D.C.: Department of the Army, 1973) pp. 257-258

[4] U.S. Special Operations Command, Publication 1, *Special Operations in Peace and War*, p. 5-6

[5] B.H.Liddell-Hart, *Strategy*, (New York: Meridian, 1991), p.4

[6] J.F.C. Fuller, *The Conduct of War 1789-1961* (London: Eyre and Spottiswoode, 1961) pp.242-243

[7] B.H.Liddell-Hart, *Strategy*, p.146

[8] James J. Schneiber, *Cybershock: Cybernetic Paralysis as a New Form of Warfare*, (Fort Leavenworth, Kansas: School of Advanced Military Studies, U.S. Army Command and General Staff College, June 1995)

[9] Brigadier General James C. Crockett comments as attache' to Russia in 1948 from Brigadier General S.L.A. Marshall, *The Soldier's Load and the Mobility of a Nation*, p. 63.

Bibliography

Arnold, James R., *Tet Offensive 1968, Turning Point in Vietnam*. London: Reed Consumer Books for Osprey Publishing LTD, 1990.

Bible, King James; AV 1611

Black, Robert W. *Rangers in WWII*. New York: St. Martin's Press Mass Market paperbacks, 1993.

Booth, Michael, Spencer, Duncan, *Paratrooper : The Life of Gen. James M. Gavin* http://www.amazon.com/exec/obidos/ASIN/0671732269/qid=942424311/sr=1-26/002-6268413-7128258

Bowden, Mark *Blackhawk down!: A story of modern war*. New York: Penguin Putnam books, 1999 and 2000. http://www.amazon.com/exec/obidos/search-handle-form/103-6593991-3811860

Bruer, William B., *Drop Zone Sicily: Allied Airborne Strike, July 1943*. Novato, California: Presidio Press paperback, 1997.

Bundeswehr, organizational document received from the German liaison office, Fort Leavenworth, Kansas: USACGSC, February 1996.

Cameron, Robert R. U.S. Army Armor Center Chief Historian, *American Tank Development during the old war: maintaining the edge or just getting by?*. Fort Knox, Kentucky: U.S. Army Armor magazine, July-August 1998.

Chadwick, Timothy B., Command Sergeant Major, *Death of the Combat Engineer Vehicle*. Fort Leonard-Wood, Missouri: U.S. Army Engineer magazine, December 1996. http://www.wood.army.mil/ENGRMAG/PB5964/perview.htm

Clancy, Tom, *Armored Cav, A Guided Tour of an Armored Cavalry Regiment*. New York, Berkley Books for Jack Ryan Limited Partnership, November 1994

Coleman, J.D. Lieutenant Colonel (Retired USA), *Choppers*. New York: St. Martin's paperbacks, 1988

Crist, Stan C. *Too late the XM8?*. Fort Knox, Kentucky: U.S. Army Armor magazine, January-February 1997, pp.16-19

Crist, Stan C. *The Case for the Airmobile, Amphibious Scout Vehicle*. Fort Knox, Kentucky: U.S. Army Armor magazine, July-August 1999, pp. 30-32.

DePuy, William E. General USA (Retired USA), *Infantry Combat*. Columbus, Georgia: U.S.Army Infantry magazine Mar-Apr 1990.

Dunstan, Simon, Sarson, Peter, Foss, Christopher, *Scorpion: Reconnaissance vehicle 1972-1994*. Osprey Military Books; printed by Reed Consumer Books, 1995.

English, John, Gudmundsson, Bruce I., *On infantry*. Westport, CT: Greenwood Publishing Group; Praeger press, 1994

Ferry, Charles P. Captain, *Mogadishu, October 1993: Personal Account of a Rifle Company XO*. Columbus, Georgia: U.S. Army Infantry magazine, Sept-October 1994, pp. 28-29.

Foley, Thomas C., Major General, Commander, U.S. Armor Center, *Commander's Hatch* column, Fort Knox, Kentucky: U.S. Army Armor Magazine, July-August 1990.

Franz, Wallace P., *Airmechanization: The Next Generation*. Fort Leavenworth, Kansas: Military Review, February 1992; pp. 59-64.
Foss, Christopher F. and Cullen, Timothy, *Jane's Armour and Artillery 98-99*. Coulsdon, UK: Jane's Information Group Ltd., 1998
Foss, Christopher F. , *Jane's Tank Recognition Guide*. Glasgow, UK: HarperCollins, 1996.
Galvin, John, R. (General, USA, ret.), **Air Assault**: *The development of Airmobile Warfare,* New York, NY: Hawthorne Books, 1969
Gameros, Charles W. Jr., 1st Lieutenant, *Scout HMMWVs and Bradley CFVs*. Fort Knox, Kentucky: Armor Magazine, September-October 1991, p.22.
Ganz, Harding A. *The 11th Panzers in Defense, 1944*. Fort Knox, Kentucky: Armor Magazine,March-April 1994, p.33.
Gavin, James M., Lieutenant General, *Airborne Warfare*. New York: Infantry Press, 1947, reprinted Battery Press, Nashville, Tennessee.
Grau, Lester W., Lieutenant Colonel (Retired USA), *Bashing the Laser Range Finder With A Rock*. Fort Leavenworth, Kansas, U.S. Army Center for Lessons Learned [CALL] database, Foriegn Military Studies Office Branch, *http://call.army.mil/call/fmso/fmsopubs/issues/techy.htm*
Grau, Lester, Lieutenant Colonel (Retired, USA), *Battle for Grozny: Russian Armored Vehiclevulnerability in Chechnya*. Fort Leavenworth, Kansas, U.S. Army Center for Lessons learned [CALL] database; FMSO branch *http://call.army.mil/call/fmso/fmsopubs/issues/rusav/rusav.htm*
Grow, Robert W. Major General (Retired, USA), *The Lean Years*, Armor Magazine, January-February 1987, p. 23.
Gurney, Gene, *A pictorial History of The United States Army, in War and Peace, from Colonial   Times to Vietnam*, New York: Bonanza Books, 1966.
Hammond, Kevin J. CPT and Sherman, Frank, CPT *Sheridans in Panama*. Fort Knox, Kentucky; U.S. Army Armor magazine, March-April 1990, p.8.
Hogg, Ian *Great Battles of World War 2*. New York: Double Day, 1987.
Hornbeck, Paul A. *The Wheel versus Track Dilemma*. Fort Knox, Kentucky: U.S. Army Armor magazine, March-April 1998, pp. 33-34.
International Institute for Strategic Studies, *The Military Balance 1995-1996*. London: Institute for Strategic Studies.
Jackson, P, *Jane's All the World's Aircraft 1996*. Couldsdon, UK: Jane's Information Group Ltd., 1996.
Jarnot, Charles A. Major, USA, *Air-Mech-Strike*. Arlington, Virginia: Army Magazine, January 2000, pp. 23-26.
Keegan, John, *Second World War.* New York: Viking Penguin Books, 1989.
Kazmierski, Michael, Colonel (Major), *United States Army Power Projection in the 21st Century: the conventional Airborne forces must be modernized to meet the Army Chief of Staff's strategic force requirements and the Nation's future threats*. Fort Leavenworth, Kansas: U.S. Army Command and General Staff College Master's Thesis, Defense Technical Information Center File # AD A226 216. *http://www.geocities.com/Pentagon/Quarters/2116/airbornetoc.htm*

Loughlin, Don Captain, (Retired USMC), *Sayonara AGS!, Sayonara scout?, Sayonara Armor?.* Fort Knox, Kentucky: U.S. Army Armor magazine, July-August 1998.

Lind, William S., Keith Nightengale, Keith Colonel (USA), John F. Schmitt, John F. Captain(USMC), Sutton, Joseph W. Colonel (USA), and Wilson, Gary I. Lieutenant Colonel (USMCR), *The Changing Face of War: Into the Fourth Generation.*", Quantico, Virginia: Marine Corps Gazette, October 1989, pp. 22-26.

Lock, John D., Lieutenant Colonel, USA, *To Fight with Intrepidity: The Complete History of U.S. Army Rangers 1622 to present.* New York: Simon & Schuster Pocket Books, 1998.

MacGregor, Douglas C. Colonel, USA, *Breaking the Phalanx: A New Design for Land power in the 21st Century*, Westport, CT: Praeger, 1997.

McDonald, Bradley N., Major USA, *Javelin.* Columbus, Georgia: U.S. Army Infantry Magazine, September-December 1998, p.6.

Macksey, Kenneth, *Guderian: Panzer General.* London: Greenhill Books Inc., 1992.

Mettler, Wolfgang, Lieutenant Colonel, German Army, *The German Airborne Anti-tank Battalion*, Columbus, Georgia: U.S. Army Infantry Magazine, January-February 1995.

Morris, James M., *History of the U.S. Army, Revised and Updated.* Greenwich, Connecticut: Brompton Books Corp, 1997

Naylor, Sean, *Farewell Sheridan!.* Army Times, September 23, 1996, pp. 14-16. Ogorkiewicz, Richard M., *Airborne Armor.* Fort Leavenworth, Kansas; U.S. Army Military Review, August 1959.

Patton, George S. *War as I knew it.* New York, Houghton Mifflin Company, 1995.

Peters, Ralph, Lieutenant Colonel (Retired, USA), *Our Soldiers, their Cities.* Fort Leavenworth, Kansas: U.S. Army Parameters magazine, Spring 1996. pp. 43-50. *http://carlisle-www.army.mil/usawc*/Parameters/96spring/peters.htm

Schaeffer; Francis C., *Escape from Reason*, Downer's Grove, Illinois: Inter-Varsity Press, 1983.

Sheehan, Neil, *Bright Shining Lie.* New York: Vintage Books, 1989.

Sherman, Frank, Major USA, *Operation Just Cause: the Armor-Infantry Team in the close fight.* Fort Knox, Kentucky: U.S. Army Armor magazine, September-October 1996, pp. 34-35

Shinseki, Eric K., Chief of Staff, General Speech to AUSA, October 1999, Fort Bliss, Texas: U.S. Army Air Defense Artillery magazine online: *http://147.71.210/adamag/Oct99/CofS.htm*

Sillary, Bob, *The Plane that Drove,* Popular Science magazine, March 2000, pp. 72-74.

Simpkin, Richard E. Brigadier General, *Race To The Swift, Thoughts on Twenty-First Century Warfare.* London: Brassey's Defence Publishers Ltd., 1987

Smiley, Lloyd, *Veteran Recounts WWII*, Great Falls, Montana: Great Falls Tribune, 4 April 1999.

Sparks, Michael L., *Crisis of Confidence in Armor?.* Harmon, Jodie (future tank art work) Fort Knox, Kentucky, U.S. Army Armor magazine, March-April 1998

Sparks, Michael and Harmon, Jodie (cover art work) *M113s maximize mechanized infantry mobility and firepower in contingency ops.* Fort Knox, Kentucky: U.S. Army Armor magazine, January-February 1995

Sparks, Michael L., *Improving Light Force Firepower with HMMWV mounted Recoilless Rifles.* Harmon, Jodie (HMMWV-106mm RR art work) Fort Knox, Kentucky: U.S. Army Armor magazine, November-December 1995 Sun Tzu, Introduction by Griffith, Samuel B.; *The Art of War.* Oxford University Press, 1984.

Thompson, Kris P. Lieutenant Colonel USA, *Trends in Mounted Warfare, Part III: Korea, Vietnan and Desert Storm.* Fort Knox, Kentucky: U.S. Army Armor magazine, July/August 1998.

Thompson, Julian C., Brigadier General, United Kingdom; *No Picnic: 3 Commando Brigade in the South Atlantic : 1982* http://www.amazon.com/exec/obidos/ASIN/0006370136/qid=953744929/sr=1-6/103-6593991-3811860

Thompson, Sir Robert, *War in Peace: conventional and guerrilla warfare since1945.* New York, Harmony Books, 1981. Time-Life Books, New Face of War editors, *Sky Soldiers. New York: Time-Life Books, 1994* http://www.amazon.com/exec/obidos/ASIN/0809486121/qid=942422138/sr=1-6/002-6268413-7128258

Toffler, Alvin C. and Heidi, *War and Anti-War: Making Sense of Today's Global Chaos.* New York: Warner Mass Market Paperback, 1995. http://www.amazon.com/exec/obidos/ASIN/0446602590/qid=953747016/sr=1-3/103-6593991-3811860

Tucker, Francis C., Lieutenant General, British Army, *The Pattern of War* (Washington D.C.: Washington D.C.: U.S. Government Printing Office 623-279/10260 USMC HQ, 1989

Tugwell, Maurice A.J., Brigadier General, British Army, *Is the day of the Paratrooper over?.* Fort Leavenworth, Kansas: U.S. Army Military Review magazine, March 1977, pp. 75-83 http://www.geocities.com/Pentagon/Quarters/2116/dayofparatrooper.htm

Van Crevald, Martin, *The Transformation of War.* New York: Free Press, 1991. http://www.amazon.com/exec/obidos/ASIN/0029331552/qid=954048760/sr=1-4/104-9565611-2862801

Walker, Jamie, *Going Off-Road with the Flyer.* Quantico, Virginia: Leatherneck, June 1999,p. 24.

Wilson, Peter A. and Gordon IV, Charles *Aero-motorization: pathway to the Army of 2010.* Carlisle, Pennsylvania: U.S. Army War College, Strategic Studies Institute Report, 1998, pp.9 http://carlisle-www.army.mil/usassi/ssipubs/pubs98/aeromotr/aeromotr.pdf

Zumbro, Ralph, *Iron chariots.* New York; Simon & Schuster Pocket Books, 1998, pp. 147-174

Zumbro, Ralph, *Tank Aces.* New York; Simon & Schuster Pocket Books, 1998, pp. 260-261

TECHNICAL REFERENCES

General references
Metric conversion charts/programs
http://www.worldwidemetric.com/metcal.htm

http://www.french-property.com/ref/convert.htm
U.S. Army, *Weapon Systems 1997 Fact Book*, Washintgon D.C.: GPO, 1997.

Ground vehicles
U.S. Army Science Board/DARPA report excerpt on Future Ground Combat Systems (FGCS) *http://www.darpa.mil/darpatech99/Presentations/Scripts/TTO/freeman.txt*
U.S. Army TACOM M113A3 FOV Fact Book *http://www.tacom.army.mil/dsa/pmtaws/m113*
United Defense M113A3 brochures, 1995, 1999
http://www.m113.com/m113a3ch.html
United Defense MTVL brochure, 1999
United Defense M8 armored Gun system brochure, 1999 M8 *Buford* or *Ridgway* Armored Gun System: *http://www.geocities.com/Pentagon/Quarters/2116/armored.htm* Ma*k Wiesel 1 and 2 brochure, 1999 http://www.army-technology.com/contractors/armoured/mak/index.html#mak3*
U.S. Caterpiller/German MaK Maschinebau GmbH (Wiesel Airborne light tank) Army Technology (U.K.) web site Wiesel family small AFVs *http://www.army-technology.com/contractors/armoured/mak/index.html#mak3* Piranha LAV family large wheeled AFVs *http://www.fas.org/man/dod-101/sys/land/row/piranha.htm http://www.army-technology.com/projects/piranha/index.html http://www.army-technology.com/projects/piranha/specs.html* Cobra small wheeled AFVs*http://www.army-technology.com/projects/cobra/index.html* Future Combat System web site (Fort Knox, Kentucky, U.S. Army Armor Center and School *http://knox-www.army.mil/center/dfd/FCV.htm*
U.S. Army Fort Knox, Kentucky Armor Center and School *Tracks versus wheels study http://knox-www.army.mil/center/dfd/WVTart.htm3*
U.S. Army TRADOC *Wheeled vs Tracked Vehicle Study Final Report*, (ACN 070846) March 1985 Studies and Analysis Activity
LeVan, Mel E., *A Comparison of the Mobility and Stability of Tracked and Wheeled Vehicles*
Bethesda, Maryland: Marine Corps Operations Analysis Group, Center for Naval Analysis
CRM 87-152/August 1987
Robert F. Unger, *MS Mobility Analysis for the TRADOC Wheeled Versus Tracked Vehicle Study*
Geotechnical Laboratory, DA Waterways Experiment Station COE, Vicksburg, Technical
Report GL-88-18, Sep 1988,
Willoughby, William; Jones, Randolph; Cothren, David; Moore, Dennis W.; Rogillio, David M. Unger, Robert F. Prickett, Terri L. *U.S. Army Wheeled Versus Tracked*
Vehicle Mobility Performance Test Program Technical Report GL-91-2, Report 1 Mobility in Slippery Soils and Across Gaps. Vicksburg, Mississippi: Geotechnical Laboratory, DA Waterways Experiment Station COE

U.S. Army Tank and Automotive COMmand (TACOM) briefing to industry, *Wheeled and Tracked Vehicles.* Jan 2000.

Air vehicles
Air Force Technology (U.K.) web site: Lockheed-Martin C-130J *Hercules*
 http://194.200.85.145/projects/hercules/specs.html
McDonnell-Douglas C-17 brochure and letter to 1st TSG (A), December 5, 1994
Military Traffic Management Command Air Deployability Engineering: C-5, C-17, C130, C-141B
 http://www.tea.army.mil/dpe/Aircraft.htm#C130
U.S. Government DOT web page: Civil Reserve Air Fleet (CRAF)
http://www.rspa.dot.gov/oet/enclos1_899.html
Boeing web page on the 747
http://www.boeing.com/bck_html/Boe747F-Facts.html
TM 1-1520-237-10 Average of Performance Charts for U.S. Army helicopters FM 1-113 Utility/Cargo Helicopters
http://155.217.58.58/cgi-bin/atdl.dll/fm/1-113/CH1.HTM
Survivability data
http://155.217.58.58/cgi-bin/atdl.dll/fm/1-113/AG.HTM
http://155.217.58.58/cgi-bin/atdl.dll/fm/1-113/AA.HTM
UH-60L USAF aircraft loading factors
http://155.217.58.58/cgi-bin/atdl.dll/fm/1-113/A-8.GIF
CH-47D USAF aircraft planning factors—C-17 is a no-go!
http://155.217.58.58/cgi-bin/atdl.dll/fm/1-113/A-8.GIF
CNN footage showing Israeli AH-1 and AH-64s firing at Hezbollah positions in daylight, 19 April 1996.

Weapons
Precision Guided Munitions, signature-less Russian ATGMs; 1st Special Warfare Training Group (A), Fort Bragg, NC video training tape for Career Management Field 18B Soldiers **Tactics, Techniques, Procedures Field Manuals**
Field Manual 7-7 *The Mechanized Infantry Platoon and Squad (APC).* Washington D.C.: U.S. Government Printing Office, U.S. Army HQ, 15 Marc 1985
http://155.217.58.58/cgi-bin/atdl.dll/fm/7-7/toc.htm
FM 17-33, *The Armored Battalion*, September 18, 1942
Field Manual 17-98, *The Scout Platoon.* Washington D.C.: U.S. Government Printing Office, US Army HQ, 1994.
Field Manual 90-4 *Air Assault operations.* Washington D.C.: U.S. Government Printing Office, U.S. Army HQ, March 1987)
Field Manual 90-10-1 *Infantryan's guide to Combat in built-up areas*
Washington D.C.: U.S. Government Printing Office, U.S. Army HQ, 1993
http://www.adtdl.army.mil/cgi-bin/atdl.dll/fm/90-10-1/default.htm
FM 90-26 *Airborne operations.* Washington D.C.: U.S. Government Printing Office, U.S. Army HQ, 1994
http://www.adtdl.army.mil/cgi-bin/atdl.dll/fm/90-26/default.htm
U.S. Army Student Text (ST) 100-3, *Battle Book.* Fort Leavenworth, KS: U.S. Army

Command and General Staff College [USACGSC], 1995.
Field Manual 100-2-3, *The Soviet Army: Troops, Organization and Equipment*
Washington, DC: U.S. Government Printing Office [GPO], 1991.
FM 100-5, *Operations*. Washington, DC: GPO, 1993

INTERNET SOURCES
11th ACR official web site
   http://www.11thcavnam.com/history.htm
Bowden, Mark, *Blackhawk Down!* web site based on Philadelphia Enquirer newspaper series
   http://www.philly.com/packages/somalia/ask/ask12.asp
Brueggeman, Gary, *The Roman Army:*
   http://www.geocities.com/Athens/Oracle/6622/legions.html
German Airborne's WWII *Air*-Mech-Strike combination pictures:
   http://www.geocities.com/Pentagon/Quarters/2116/hptll..htm
Korean war web site
   http://rt66.com/~korteng/index.htm
Pike, John, *Federation of American Scientists Land warfare systems web site:*
   http://www.fas.org/man/dod-101/sys/land/index.html

CORRESPONDENCE/INTERVIEWS
Anderson, Robert, Special Markets Director, Polaris Industries, telephone interview with LTC Richard Liebert , 28 Nov 1999.
Ferrell, Ronald Chief Warrant Officer 4, U.S. Army Aviation Center resident expert on RAH-66 capabilities, interview by Major Charles Jarnot, Fort Rucker, Alabama, 6 April 1996.
Garrett, Steve, USARNG, author of FM 7-7J section on Bradley Stingray EOCM system, personal interview with Michael Sparks, December 5, 1999
Grusonik, Bill, United Defense LP M113A3 Engineer; March 1, 2000, e-mail to Michael Sparks on exact M113A3 weights
Hemberger, Joe, Staff Sergeant, USAF, *ATVs in USAF use*. ATV Safety Trainer, Minnesota Air National Guard, e-mail to Lieutenant Colonel Richard Liebert, 24 February 2000.
Holman, Eric Master Sergeant, USAF, ATV expert, Lackland AFB, 30 August 1999, e-mail to LTC Richard Liebert (eric.holman@lackland.af.mil)
Moore, Harold G., Lieutenant General (Retired, USA); telephone interview with Michael Sparks; January 15, 2000
Parker, Bob, Flyer Group Inc, 17 Dec 1999 telephone interview with LTC Richard Liebert and Flyer Group literature, Oct ober 1999; (flyergroup@aol.com)
Shelton, Henry H., General, USA, CINCUSSOCOM, letter to Michael Sparks, March 14, 1997 in response to his suggestion that SOF acquire M113A3 typearmored vehicles to be a "Ground spectre" fire support vehicle for direct action missions
Steinhauer, Dale R., resident expert on Middle East wars, interview by Major Charles Jarnot, 6 April 1996, Army Knowledge Network, USACGSC, Fort Leavenworth, KS, 6 April 1996.

Wass de Czege, Huba, Brigadier General (Retired, USA) lecture on advanced warfighting in the early 21st century (Fort Leavenworth, KS: USACGSC, 10 January 1996).

OTHER MEDIA

CNN report on crossing the Sava River in Bosnia, 12 January 1995.

Appendix A:

Military Abbreviations and Acronyms

A-10	"*Warthog*" attack aircraft;
Abrams	M1 Main Battle Tank
AFVs	Armored Fighting Vehicles
AH-64	*Apache* attack helicopters
AASLT	Air Assault by helicopters
ABN	Airborne operations
AMS	*Air*-Mech-Strike! concept
AK-47s	*Kalishikov* automatic rifles; AKMs
ATBs	All-Terrain Bicycles
ATACs	All-Terrain All-purpose Carts
ATVs	All-Terrain Vehicles
Applique armor	Bolt-on armor panels to improve vehicle protection
AR/AAV	Armored Reconnaissance/Air Assault Vehicle or M551 *Sheridan* light tank
ACR	Armored Cavalry Regiments
Armored cars	Wheeled armored platforms
AAN	Army After Next
ARVN	Army of the Republic of Viet Nam Soldiers
Asymmetric warfare	Fighting against an enemies weaknesses with your strengths
ATGMs	Anti-Tank Guided Missiles
battle groups	1950s *Pentomic* Army formation to fight on dispersed nuclear battlefield
Blackhawk	UH-60 helicopters
Biological warfare	Use of disease vectors to spread disease/illness
BMD	8-ton Russian armored fighting vehicle
BOS	Battlefield Operating Systems
body armor	Vest and helmet combination to protect the Soldier
Bradley	5-star Army General, Omar N. Bradley
BFV	*Bradley* Fighting Vehicle
RAAWS	*Carl Gustav*, M3 Ranger Anti-Armor Assault Weapon System
CH-47D/F	Heavy-lift *Chinook* helicopters
Camouflage	Face paint or vehicle strips to evade enemy detection
Cavalry	Armed reconnaissance forces
Close Air Support	Usually Fixed-wing aircraft firing weaponry for ground maneuver
combined-arms	Achieving synergistic effects against enemies via various type unit capabilities
Churchill, Winston S.	Prime Minister of England and military nnovator
Darby, William O., BG	Leader of 1st Ranger battalion in WWII

Damon, Samuel	Fictional war hero in novel *Once an Eagle* who is combat focused yet puts men before self and career
Depuy, William, GEN	Army General who helped create TRADOC
Digitization	Using computerized communications to get greater awareness
Dragoons	Armed, mounted Soldiers
EOCM	Electro Optical Counter Measures, laser targeting means
EFOGM	Enhanced Fiber Optic Guided Missile
Fallschirmjaegers	German Paratroopers
FRIES	Fast Rope Insertion Extraction System
FLIR	Forward Looking InfaRed or "thermal" sights
Flyer 21	3-ton large All-Terrain-Vehicle
Gavin, James M., LTG	Lieutenant General, Airborne Army theorist
Gliders	Unpowered assault aircraft
GPS	Gobal Positioning System
Gunshields	A shield on the end of weapons to protect against enemy fire
HMMWV	High Mobility Multi-purpose Wheeled Vehicle
Javelin ATGM	Fire/Forget infrared self-guiding Anti-Tank Guided Missile
jeep, M151	4x4 All Terrain Vehicle
Kosovo	Province of Serbia dominated by ethnic Albanians
Lasers	Amplified light beams
Liddell-Hart, Basil Sir	Great military theorist of the indirect approach
Lind, William S.	Maneuver warfare theorist
Lee, Robert E.	General of the CSA during U.S. Civil war
LOSAT	Line of Sight Anti-Tank missile
Maneuver warfare	Concept of collapsing enemy cohesion to defeat him
MacArthur, Douglas, Gen	General of the Armies; former CSA in the 1930s
machine guns	Rapid fire, usually belt-fed small arms weapons
M1A2	*Abrams* Main Battle Tanks
M113A3	*Gavin* Fighting Vehicle
M114	Scout vehicle used in the 1960s
M4	*Sherman* medium tank or 5.56mm carbine
M8	17-ton Armored Gun System or WWII Greyhound armored car
M16	5.56mm automatic rifle
M50 *Ontos*	armored vehicle with 6 x 106mm recoilless rifles
M551 *Sheridan*	AR/AAVs
Moore, "Hal" Harold G.	Lieutenant General; led 1st BN, 1st Cav at LZ X-Ray
Mortars	Smooth-bore tube which rounds are fire indirectly
mechanical advantage	Using physical means to gain positional advantage
nuclear weapons	fission or fusion based explosive devices
NVA	North Vietnam Army
OH-58D	*Kiowa Warrior* scout/attack helicopter
Desert Storm	War against Iraq in 1990-91 to evict it from Kuwait

Paratroopers	Soldiers trained to fight from the sky
Patton, George S., Gen	Armor General officer and armored warfare theorist
Patton, George S. Jr.	Son of General who led 11th ACR in Vietnam
Pentomic Army	Army organization in the 1950s to stay dispersed on nuclear battlefields
Phalanx, ancient	Formation of Soldiers locking shields together for armored protection
Pinzgauer	4 and 6 wheel large ATVs
Rangers	Elite Army light infantry
Ridgway, Matthew B.	General and CSA; Airborne Commander in WWII, led UN forces in Korea
Reactive armor	Armor tiles that explode outward to deflect anti-armor warheads
Recoilless Rifles	106mm and 84mm weapons using exhaust gases escaping to the rear
RMA	Revolution in Military Affairs
Rifle grenades	Grenades launched by gas or trapped bullets of rifles
Rommel, Erwin	German WWII Field Marshall noted for tactical genius
Roosevelt, Theodore	"Teddy"; U.S. President;
RPGs	Rocket Propelled Grenades
"*Scud*"	Surface to Surface Missiles
SEAD	Suppression of Enemy Air Defenses
shock action	Stunning an enemy with overwhelming force and/or positional advantage
Simpkin, Richard	*Air*-Mech warfare theorist
Smokescreens	Obscurrants to cover forces from enemy fire/observation
Snipers	Skilled marksmen that can hit targets precisely with small arms
Somalia	War-torn East African country
Space-based targeting	Using satellites to target military forces
Speed marching	Marching at high foot speeds to get positional advantage
Stuart, JEB	CSA Cavalry leader
T-10C	Round static-line parachute
Task Force *Hawk*	Aviation and Airborne Task Force in Tirana, Albania
Technotactical	The technical aspects of war with tactical implications
Toffler, Alvin/Heidi	Futurist authors and husband/wife team
TOW ATGM	Long-range, heavy Wire-Guided Anti-tank missiles
Tucker, Francis LTG	British future war theorist
VC	"Viet Cong" Guerrilla forces that fought during 1958-75
VDV	Russian acronym for Desant or Airborne troops
WMD	Weapons of Mass Destruction
Zumbro, Ralph	*Air*-Mech theorist and Vietnam combat armor veteran

Bell / Boeing Quad Tilt-Rotor

High Cost FTR Concept

Up Scaled V-22 (+)
20+ Ton Lifting Tilt-Rotor
High Cost FTR Concept

Good delivery concept depicted: Bomb-Bay Door Delivery to Winch FCS vertically Through Un-Even Broken Terrain avoiding airlanding risks in open-area LZs; concept applicable to other aircraft types!

Appendix B
Air-Mech-Strike Technology Concepts

CH-53E-S "SPEED STALLION"
Low-Cost Compound FTR Concept

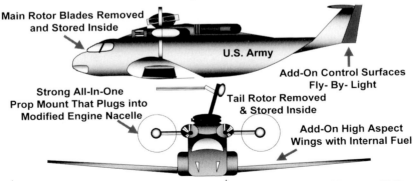

- Main Rotor Blades Removed and Stored Inside
- U.S. Army
- Add-On Control Surfaces Fly-By-Light
- Strong All-In-One Prop Mount That Plugs into Modified Engine Nacelle
- Tail Rotor Removed & Stored Inside
- Add-On High Aspect Wings with Internal Fuel

★ No Wear on Rotor Drive Train.
★ Internal "Y" Shaft to #3 Engine.
★ 250 KIAS Cruise = C-130 Speed.
★ Self-Deploy Asia/Europe 48 Hrs
★ Wings Dropped & Rotor Blades Installed Upon Arrival in Theater

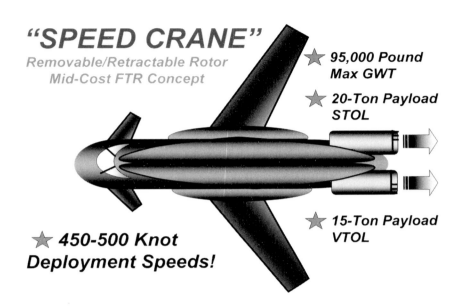

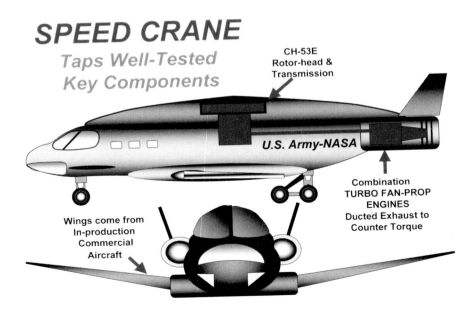

SPEED CRANE
Switch-Blade-Rotor (SBR) Concept

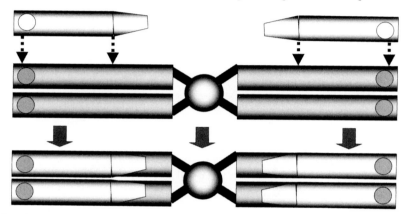

Stiff Lower Blade Sections held on short root arms.
Top Blade Sections held by large circular pins.

SPEED CRANE
Switch-Blade-Rotor
Retractable Rotor FTR Concept

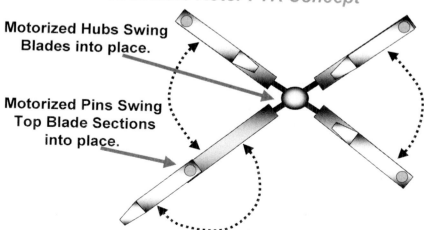

Motorized Hubs Swing Blades into place.

Motorized Pins Swing Top Blade Sections into place.

SPEED CRANE
Removable/Retractable Rotor
FTR Concept

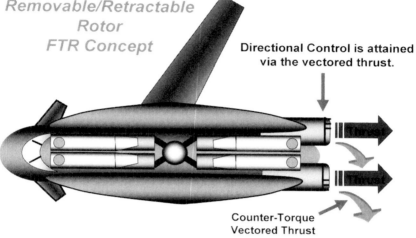

Directional Control is attained via the vectored thrust.

Counter-Torque Vectored Thrust

SPEED CRANE
Large Pod

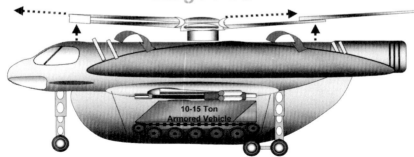

10-15 Ton Armored Vehicle

Large Pod/Basket for oversize loads. Cables support metal floor for load, shell carries no structure load, only designed to reduce drag.

SPEED CRANE
Medium Pod

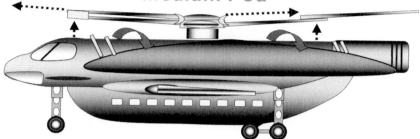

Medium Pod for Personnel or light cargo.
Each pod is tailored to each service need. "Modularization"
of the Pod concept separates the multi-service debate
on cabin size, doors, ramps and seating capacity. All
Army units can have pods for training, pre-loaded for
operational readiness. Glider pods for stealth assaults.

SPEED CRANE
Small Pod

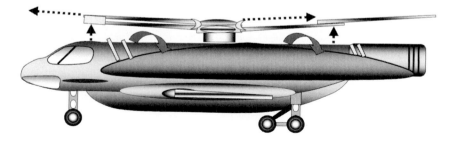

Small Pod ideal for extra fuel for self-deployments.
Electronic Warfare, Anti-Submarine Warfare, C/SAR
and even Stand-off Attack Missions.

SPEED CRANE
Stripped For Heavy Lift

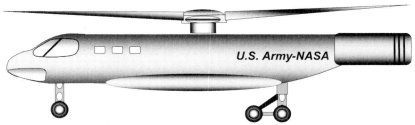

U.S. Army-NASA

Once Self-Deployed in Theater,
Speed Crane can have the wings dropped, Rotor-Head
Cover dropped and conventional one-piece blades
installed to maximize lift efficiency.
Small Stub Wings could be added to increase speed.

SPEED CRANE
Retractable/Detachable Rotor FTR Concept

U.S. Army-NASA

- Low-Risk Development via CH-53E Transmission & Rotorhead and commercial aircraft wings.
- Low-Cost compared to Tilt-Rotor, fewer rotors, fewer transmissions, fewer moving parts under-load.
- Low Maintenance via Part Time Rotor.
- Better Stealth / Signature Management, No moving rotors exposed.
- One Day Self Deployment Speeds, 30% faster than Tilt-Rotor.
- Modular Pods = 100% Multi-Service Needs.
- Glider pods = Airborne/SOF special mission needs

20-Ton Gun-Missile FCS

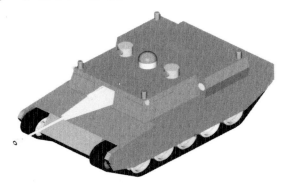

20-Ton
35mm ETC Gun & LOSAT
Medium cannon (Crew)
14.5 mm AP (Sides & Turret)
CE/KE APS (With KE standoff)

15-Ton "*AEGIS*" FCS

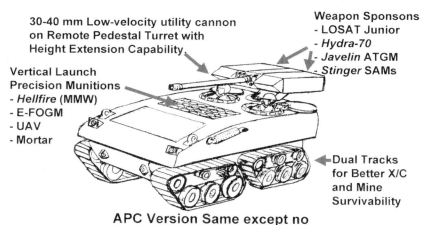

30-40 mm Low-velocity utility cannon on Remote Pedestal Turret with Height Extension Capability

Weapon Sponsons
- LOSAT Junior
- *Hydra-70*
- *Javelin* ATGM
- *Stinger* SAMs

Vertical Launch Precision Munitions
- *Hellfire* (MMW)
- E-FOGM
- UAV
- Mortar

Dual Tracks for Better X/C and Mine Survivability

APC Version Same except no Vertical Launch Precision Munitions

T-MARS
(Trailer-Multiple Artillery Rocket System)

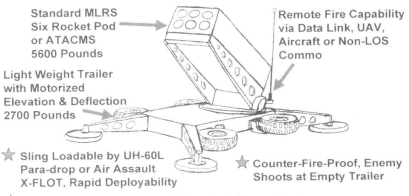

Standard MLRS Six Rocket Pod or ATACMS 5600 Pounds

Remote Fire Capability via Data Link, UAV, Aircraft or Non-LOS Commo

Light Weight Trailer with Motorized Elevation & Deflection 2700 Pounds

★ Sling Loadable by UH-60L Para-drop or Air Assault X-FLOT, Rapid Deployability

★ Counter-Fire-Proof, Enemy Shoots at Empty Trailer

★ Max Mass Fires - All T-MARS Ready to Fire without Expensive/Heavy MLRS Launch Vehicle

FCS High-Velocity Gun 15-20 ton

UPGRADED M8 AGS　　　　**NEW DESIGN FROM SCRATCH**

Possible features

*Band tracks
*Electric drive
*Active protection systems

*Composite construction
*Stealth camouflage strips
*Engineer blades/attachments

FCS LOSAT
External Gun/Missile
10-20 tons

Index

A
A-10 113
AAN 238, 240, 271
Abrams 10, 52, 79, 100, 109, 151, 155, 156, 157, 164, 176, 188, 205, 220
AFVs 7, 44, 48, 51, 52, 56, 63, 65, 66, 67, 80, 81, 82, 83, 86, 88, 124, 129, 142, 147, 150, 154, 157, 158, 159, 162, 163, 166, 168, 192, 194, 221, 235
AH-64 86, 95, 96, 100, 101, 113, 190, 219
Air Assault doctrine 107
Air-Mech-Strike concept 60, 80, 82, 83, 85, 90, 93, 94, 95, 96, 106, 111, 112, 114, 128, 160, 179, 233, 247, 252, 277
Airborne operations 39, 43, 67, 68, 121, 277
Ap Bac *(Battle of)* 49, 50, 67
Applique armor 205
AR/AAV 13, 45, 47, 132
armored cars 41, 78, 124, 126, 136, 140, 143, 145, 146, 148, 149, 156, 157, 161, 227, 244
Armored Cavalry Regiments 48
ARVN 49, 66, 67
asymmetric 34, 149, 215, 226, 233, 279
ATACs 161, 162, 163
ATBs 161, 162, 163, 171, 180, 183, 189
ATGMs 69, 94, 109, 163, 164, 176, 180, 199, 201
ATVs 87, 159, 160, 161, 162, 163, 164, 168, 169, 170, 172, 174, 179, 180, 182, 183, 189, 190, 200, 217, 235

B
BMD 54, 56, 58, 72, 78, 83
body armor 175, 180
BOS 165
Bradley Fighting Vehicle 99, 145, 156, 157, 176, 204, 218
Bradley, Omar N. 232

C
camouflage 22, 63, 180, 199
Carl Gustav 193, 194, 203
Cavalry 13, 15, 17, 32, 44, 45, 47, 48, 49, 50, 51, 52, 53, 56, 62, 66, 70, 77, 79, 168, 173, 175, 178, 179, 188, 205, 206, 234, 242, 245, 279
CH-47 7, 8, 31, 45, 46, 48, 52, 60, 78, 84, 85, 87, 90, 91, 92, 95, 107, 111, 114, 115, 121, 124, 125, 128, 137, 138, 139, 147, 155, 156, 166, 177, 182, 183, 241, 242, 244, 246, 247, 249
Churchill, Winston 76
Close Air Support 22, 36, 46, 164, 169, 182
Combined Arms 23, 234, 279

305

D
DePuy, William E. 74
digitization 9, 82, 85
Dragoons 163

E
EFOGM 82, 204
EOCM 116, 118, 158, 195, 197, 198, 201, 203

F
fallschirmjaegers 39
FLIR 86, 95, 179, 198
Flyer 21 163, 167, 175, 176, 177, 180, 181, 182, 183

G
Gavin, James M. 16, 27, 33, 43, 66, 76
gliders 168
GPS 86, 133, 164, 173, 180, 181, 199, 222
gunshields 188, 192

H
HMMWV 52, 77, 124, 145, 147, 152, 167, 168, 169, 171, 175, 176, 179, 181, 183, 188, 192, 195, 199, 203, 204
hoof speed 76

J
Javelin ATGM 150, 163, 173, 180, 181, 182, 188, 194

K
Kosovo 17, 21, 52, 59, 63, 78, 92, 96, 119, 125, 139, 175, 183, 204, 215, 217, 218, 219, 226, 227, 259, 277

L
laser 86, 87, 116, 118, 190, 197, 199, 201
Liddell-Hart, B.H. 23, 60, 220, 282
LOSAT 82, 109, 165, 195, 199, 206

M
M113A3 7, 8, 31, 52, 62, 72, 84, 86, 92, 114, 122, 125, 126, 128, 134, 136, 137, 143, 145, 148, 149, 155, 156, 157, 158, 159, 189, 192, 197, 198, 244, 245, 246
M114 48, 66
M151 Jeep 52, 118, 176
M1A2 165
M4 45, 67, 68
M50 Ontos 48
M551 44, 45, 48, 51, 59, 60, 62, 77, 118, 121, 132
M8 8, 31, 60, 72, 114, 122, 148, 157, 158, 159, 160, 162, 188, 197, 205, 241, 246, 249
machine guns 7, 37, 41, 42, 45, 49, 50, 51, 52, 76, 88, 163, 164, 172, 180, 188, 223, 227
maneuver warfare 13, 14, 28, 61, 64, 76, 77, 80, 93, 95, 127, 128, 139, 282
mechanical advantage 73, 74
Moore, Harold G. 10, 14, 66
mortars 76, 78, 109, 169, 195, 200, 203, 206, 223

N
nuclear weapons 256, 258

O
OH-58D Kiowa Warriors 152, 190
Operation Desert Storm 10, 48, 66, 89, 90, 94, 96, 113, 118, 215
Operation Just Cause 10, 71

P
Paratroopers 34, 39, 40, 42, 44, 52, 58, 59, 67, 72, 118, 125, 163, 217, 219
Patton 17, 44, 51, 66, 236
Phalanx 7, 16, 17, 20, 26, 27, 32, 33, 100, 135, 154, 215, 241, 246, 247, 252, 279, 281
pods 190, 224, 251

R
Rangers 62, 73, 118, 171, 174, 280
Recoiless rifle 84mm 193, 194
Recoiless rifle 106mm 48, 49, 52, 197, 201, 206
reactive armor 197
Revolution in Military Affairs 29, 255, 259
Ridgway, Matthew B. 201
Rommel 17, 227
RPGs 86

S
Scud 22, 60
SEAD 116, 218
Shinseki, Eric K. 13, 19, 25, 32, 64, 260, 262, 272
shock action 27, 60, 158

Simpkin, Richard 39, 65, 134
smokescreen 204
Sniper 192, 193, 194, 197, 198, 201, 203, 206
Somalia 55, 56, 72, 132, 147, 172

T
Task Force Hawk 217
Toffler, Alvin and Heidi 74
TOW ATGM 164, 188, 197, 201
Tucker, Francis 73

U
UH-60 8, 79, 82, 84, 85, 87, 90, 92, 95, 97, 100, 107, 111, 114, 115, 124, 128, 152, 155, 157, 166, 177, 182, 189, 190, 191, 200, 204, 205, 222, 223, 241, 246, 247

V
VC 50
VDV 54, 72
Vietnam 7, 8, 10, 12, 13, 15, 45, 46, 47, 48, 49, 51, 52, 55, 56, 58, 66, 67, 94, 95, 120, 121, 131, 183, 251, 257, 282

W
Wass de Czege, Huba 7, 101
WIG 263, 264, 265, 266, 268, 269, 270, 271, 272, 274

Z
Zumbro, Ralph 8, 56, 64, 66, 70, 121

Biographical Sketches of the Staff

Brigadier General David L. Grange, U.S. Army (Retired), began serving as Executive Vice President and Chief Operating Officer of the Robert R. McCormick Tribune Foundation in November 1999. He is charged with the responsibility of managing all administrative and fund-raising efforts of the RRM Foundations. He also serves as a strategic planner, coordinates the operations of the three foundations, and establishes cooperative relationships with key officials from outside organizations.

Grange came to the foundation after 30 years of service in the U.S. Army, His final assignment was as Commanding General of the 1st Infantry Division, also known as the "*Big Red One*". In that position, he served in Germany, Bosnia, Macedonia, and Kosovo. During his military career, Grange served as a Ranger, Green Beret, Aviator, infantryman, and as a member of Delta force. Assignments and conflicts took him to Vietnam, Korea, Grenada, Russia, Africa, former Warsaw Pact countries, Central and South America, and the Middle East. His awards include 2 Distinguished Service Medals , 3 Silver Stars, 2 Defense Superior Service Medals, 3 Legions of Merit, 2 Purple Hearts, Vietnam Gallantry Cross, French Ordre *National du Merite, German das Ehrenkreuz de Bundeswehr* in Gold, Combat Infantryman's Badge (CIB), Special Forces TAB, RANGER TAB, AVIATOR badge, SCUBA badge, Military Free-Fall (HAHO/HALO) JumpMaster MASTER PARACHUTIST static-line, Pathfinder badge, Air Assault badge, Expert Infantry Badge, with Oak Leaf Clusters, with Oak Leaf Clusters, ,, NATO Medal, Republic of Vietnam Campaign Medal, Kuwait Liberation Medal (Kingdom of Saudi Arabia), Kuwait Liberation Medal (Government of Kuwait).

A Long Island, New York native, born December 29, 1947, Grange holds of bachelor of science degree from North Georgia College and a Masters Degree of Public Service from Western Kentucky University. While in the military, Grange attended Infantry Officer course, U.S. Marine Corps Command and General Staff College, the British Special Air Service, and the National War College. Grange, his wife, Holly and their two sons, David and Matthew, reside in Wheaton, Illinois.

Brigadier General Huba Wass de Czege, U.S. Army (Retired), was born in Kolozsvar, Hungary on 13 August 1941. He entered the United States Military Academy, at West Point New York in 1960 and graduated in 1964 as a Second Lieutenant of Infantry. His career includes 24 months of troop service in Vietnam as a Company Commander and as a Ranger Battalion Senior Advisor. He has also served with troops in Germany. Fort Lewis, Washington and Fort Ord California in both heavy and light infantry units.

Between August 1980 and July 1982 he served as author of FM 100-5 *Operations* and as Chief of Doctrine at the U.S. Army Command and General Staff college at Fort Leavenworth Kansas. From June 1983 to May 1985 he designed and founded the initial School of Advanced Military Studies (SAMS) at the U.S. Army Command and General Staff College. Following Command of a light Infantry Brigade at Fort Ord California, he was assigned as Special Assistant to the Chief of Staff at SHAPE from June 1988 to June 1989. Brigadier General Wass de Czege became Chief of Arms

Control Branch at SHAPE Belgium in June 1989. In this capacity he provided military advice to NATO during the Conventional Forces in Europe (CFE) in Vienna. He then served as Assistant Division commander for Maneuver of the 1st Infantry Division at Fort Riley Kansas. His awards include the Silver Star, Legion of Merit (with oak Leaf cluster), Bronze Star Medal with two "V" devices (with four oak leaf clusters) Combat Infantry Badge, Senior Parachutist Badge and Ranger Tab. His Military Education includes Infantry officer basic and advanced courses, U.S. Army Command and General Staff College and the U.S. Army War College. He has a Bachelor of Science in Engineering from West Point, and Master's Degrees from Harvard in International Politics and Economics and from The U.S. Army Command and General Staff College at Fort Leavenworth in Military Arts and Sciences.

Lieutenant Colonel Richard D. Liebert USAR is a cattle rancher and CAS3 Staff Leader in the USAR. He was commissioned infantry from Purdue University, where he earned an agricultural degree. His minor was military history and he stresses that subject to his captains that he trains and coaches in the Combined Arms & Services Staff School, one of the best schools in the Army for teaching problem solving, communications skills, staff teamwork and learning about Army operations.

He served with 1-26 Infantry (Mechanized) of the 1st Infantry Division in Germany, then 1st CAV in Texas as BDE operations officer then left the Army dissatisfied with "*Hollow Army*" politics. He joined the NYARNG in '81 and met his wife, got married and has five kids. He was in the 42nd Infantry Division and was in the 1-71 Infantry (Lineage: Civil War, Spanish-American, WWII); BMO, S3-Air, S2, CEO, and company command. During this time in New York City, he was a private investigator. He then taught ROTC at St. John's University on active duty, and went into the MTARNG when his wife and him decided to leave the troubles of the cities and become farmers. He was in STARC MTARNG serving as XO of 1-163 Infantry (Mech, M113A3) and then S3 of M2A2 Bradley Fighting Vehicle after going through reorganization and NET from Mech M113A3 to BFV.

His ATV ideas come from his knowledge of military history and practical and common business sense use of Technology that allows family farmers to survive in a demanding business environment. LTC Liebert blends his '"words and plowshares" experiences regarding how new technology, tactics, weapons to better the Army.

Lieutenant Colonel Lester W. Grau, U.S. Army, Retired, is a military analyst with the Foreign Military Studies Office, Fort Leavenworth, Kansas. He received a B.A. from the University of Texas, El Paso and an M.A. from Kent State University. He is a graduate of the U.S. Army Command and General Staff College, the U.S. Army Russian Institute, the Defense Language Institute and the U.S. Air Force War College. He fought in Vietnam and served in a variety of command and staff positions in the US, Europe and Korea as an infantry officer and Foreign Area Officer. He has written extensively on Russian and Soviet military topics.

Major Charles A. Jarnot, U.S. Army was originally commissioned as an Armor officer and served as a Tank Platoon Leader before transferring to Aviation. He is a Master Army Aviator, who has commanded an Air Cavalry Troop, Attack Helicopter Company and a Heavy Lift Training Support Company. He has served as a National Training Center Observer Controller for 28 rotations, a Strategic Force Projection Planner in the Pacific Theater and has had key training assignments as an Instructor Pilot and Advanced Officer Course Small Group Instructor.

Major Jarnot has a combined-arms military education with Armor Officer Basic Course, Infantry Officer Advanced Course and resident U.S. Army Command and General Staff College. He has a Bachelor of Science Degree In Aviation Maintenance Technology from Western Michigan University in 1981, a M.A.S. in Aerospace Operations from Embry Riddle Aeronautical University in 1993 and a M.M.S. from the U.S. Army Command and General Staff College in 1996 with a thesis in *Air*-Mechanization.

He has published articles in the *Army Aviation Association of America Magazine*, *Infantry Magazine*, *Military Review* and *Army Magazine*. His first work on *Air*-Mechanization dates back to 1981 with an unpublished essay on *"Rotor-Blitz"*. Major Jarnot is from Buffalo New York and is married to the former Paula J. Barczewski. They have four children, Walter Daniel, Catherine, Stuart and Charles Elijah.

Emery E. Nelson is a noted military veteran and analyst, football coach, and father now working on his doctorate in history and was an enlisted man in the Army from '72 through '75. He was in 4/69th Armor in Mainz in the 8th Infantry Division. He is the Chief tactician in charge of heavy mortar and urban maneuver for the 1st Tactical Studies Group (Airborne). He has worked in the construction industry for several years and now owns an Elderly Care Facility and is now finishing his work on his master's degree in Anthropology at Humbolt College. Presently he is attending Columbia College and getting certified to work with Geographic Information systems.

Michael L. Sparks, a 1988 graduate of Liberty University, has a Bachelor of Science degree in history education. He leads a non-profit think-tank, the 1st Tactical Studies Group (Airborne) originally based out of Fort Bragg, North Carolina, that has field-tested and proven many equipment, offering them for no charge to the U.S. Army. He has been involved in amphibious infantry, combat engineer, and logistics, and civil affairs, Special Forces, light M113A3 mechanized infantry, and heavy BFV mechanized infantry units throughout his 19-year career. A prior service marine NCO and officer, he is a graduate of Infantry Training School, Camp Pendleton, California, Officer Candidate School Platoon Leaders Course Junior and Senior, The (officer) Basic School, Infantry Officers Course, Israeli Airborne school, Combat Lifesaver, and U.S. Army Airborne and Retention schools. His numerous articles on military excellence have been published in U.S. Army *Infantry*, *Armor*, *Air Defense Artillery Online*, *Special Warfare*, *Logistician*, the *Rucksack*, *Armed Forces Journal International*, *National Guard*, *Marine Corps Gazette*, U.S. Naval Institute *Proceedings*, *Behind the Lines*, *SOF, Mountain Bike*, Fort Bragg *POST*, Fort Benning *Bayonet* and soon *Military Review.*

Jacob W. Kipp is a senior analyst with the Foreign Military Studies Office, Fort Leavenworth, Kansas. He graduated from Shippensburg State College and received a Ph.D. from Pennsylvania State University. He has published extensively in the fields of Russian and Soviet military history and serves as the American editor of the journal European Security. He is an Adjunct Professor of History with the University of Kansas and teaches in the Russian and European Studies Program.

John Richards is the staff artist for this work. He was born December 7, 1966 in Milwaukee, Wisconsin. He graduated from Menomonee Falls High School in Menomonee Falls, Wisconsin in 1985.

In September 1985, he joined Army as 31C Radio Operator; stationed at Fort. Polk, Louisiana, two National Training Center rotations supporting DIVARTY missions against OPFOR, earned an Army Achievement Medal. From September 1988 to April 1990, he served with 12th Special Forces Group in Arlington Heights, Illinois as a Commo Team Leader and supported *Brimfrost '89* at Fort. Wainwright, Alaska. Additional duties were Soviet and threat recognition. In March 1991, he graduated from Airborne School at Fort Benning, Georgia. On May 1991-March 1992 he completed Special Forces Assessment and Selection (SFAS) at Ft. Bragg, followed by 2 months AIMC, 6 weeks of Survival, Evasion, Resistance, Escape (SERE) Survival support and 4 weeks INTAC support for USAJFKSWCS. From January 1996 to Jan 1998 was an Imagery Analyst and Intelligence Analyst for the Nevada Air National Guard; completing two rotations to Howard AFB, Panama, and one rotation to Roswell New Mexico for *Roving Sands '97*. From November 1999 to the present assigned to 82nd Airborne Division LRSD Communications section as Assistant Team Leader for LRSD HF and FM requirements; one JRTC rotation. Mr. Richards is self-taught in Russian and Spanish, working on Associates Degree when not pursuing military art projects.

Carol A. Murphy is the internet researcher and computer specialist for the staff. Carol was born and raised in New York, where she earned her certificate in Medical Assisting from The College of Staten Island. Ms. Murphy is co-director and web-designer of 1st Tactical Studies Group (Airborne), who sponsors the *Air*-Mech-Strike Study Group (Airborne). She has dedicated herself in helping 1st Tactical Studies Group (Airborne) bring their improvements, equipment and ideas to life by researching, scanning pictures, designing, and sometimes testing the ideas out on herself. She spends numerous hours typing, editing, and proofreading documents for publication in official Army journals, book publishing and web page construction.